COVID-19 & Robotics

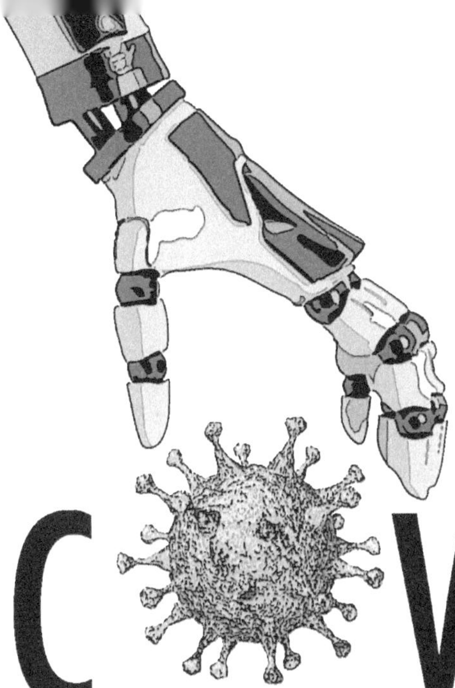

COVID-19 & ROBOTICS

Austin Mardon

Nicholas Lum, Maria Xiao, Danaeya Wolfe

and

Catherine Mardon

2020

Copyright © 2019 by Austin Mardon

All rights reserved. This book or any portion thereof may not be reproduced or used in any manner whatsoever without the express written permission of the publisher except for the use of brief quotations in a book review or scholarly journal.

First Printing: 2020

ISBN 978-1-77369-162-6

Golden Meteorite Press
103 11919 82 St NW
Edmonton, AB T5B 2W3
www.goldenmeteoritepress.com

We acknowledge the support of Canada Service Corps, TakingITGlobal and the Government of Canada in promotional materials associated with the Project.

Thank you

CONTENTS

Ch. 1: Introduction 1

Ch. 2: The Context of COVID-19 6

Ch. 3: The Burden on Humans 12

Ch. 4: The Advantages of Robotics in pandemic situations 18

Ch. 5: The Disadvantages of Robotics in pandemic situations........... 24

Ch. 6: Robotics In Retail Industries 30

Ch. 7: Robotics Automation in Food Industry 35

Ch. 8: Artificial Intelligence in COVID-19 Screening 41

Ch. 9: Artificial Intelligence in COVID-19 Research & Treatment.... 47

Ch. 10: Robots Used in the Hospital During COVID-19 53

Ch. 11: Robots in Quarantine and Nursing Homes 60

Ch. 12: Conclusion 66

References 72

Chapter 1

INTRODUCTION

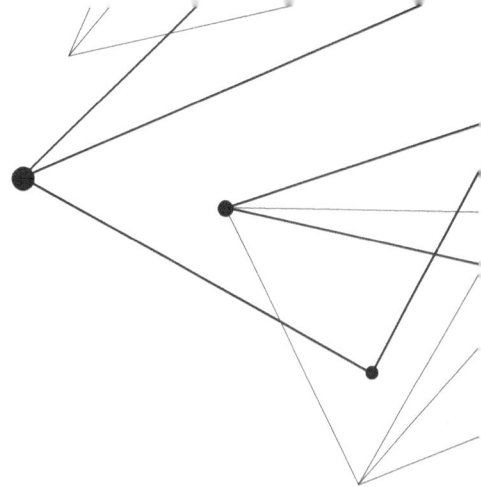

On Friday March 13th, 2020, I - like many other students - woke up to the sound of birds and the filtered light of the sun as it passed through the blinds. I checked my phone for the time: it was 9:30 AM. Just another day at McMaster University - well, not any typical day. I had a chemistry midterm later that evening, but by then I had grown accustomed to the academic rigours of a post-secondary environment. As part of my daily routine, I got up and grabbed something to eat before heading over to my desk to check my emails for the day. As I mindlessly combed through my unread messages, there was one that caught my eye: a message from our assistant dean regarding a major update on the COVID-19 situation. My heart started to race. Only a couple of days prior, some of us were joking about the "clairvoyance" one of our class facilitators provided when she said that COVID-19 might shut down the university in the next week. It is not that I thought her prediction was preposterous; the closing of other Ontario universities earlier in the same week gave an ominous foreshadowing of events to come. Additionally, the World Health Organization ([WHO], 2020) had officially declared COVID-19 as a pandemic on March 11th, 2020. I brushed off her words and tried to downplay the severity of the virus. My first year of university had been an unforgettable experience and I did not want to even think about the notion of the academic year ending so abruptly. Little did I know, the year was also about to become an unforgettable time for the rest of the world. Hesitantly, I hovered my cursor over the email, took a deep breath, and clicked.

Just like that, the university shut down indefinitely. The assistant dean forwarded a message from the university president, who stated that in-person classes for students would be ending at the end of Friday March 13th and no in-person exams would be held at the end of the term. Many of us were aware of the virus' potential to spread like a wildfire and therefore have massive implications for society, yet it still seemed so surreal. Surprisingly, my chemistry midterm still occurred as scheduled, so I did not have much time to focus on COVID-19's potential impacts. I remember calling my parents and telling them that I would be going home later that night. Everything was moving so fast and the day passed by in a blur. Once I finished my test, a weird feeling started to settle in. In retrospect, I was still numb to the situation. I said my goodbyes to my friends and peers, letting them know that I would hopefully be seeing them soon. After I grabbed a bite to eat, I made my way to the residence building to pack my bags before my parents arrived to take me back home.

In the coming days, news of other universities closing began to emerge. I learned that the province of Ontario had shut down schools in the week leading up to the university's closure. More cases started to appear in the province, country, and around the world. More money was funnelled into research. People were starting to stay home more frequently due to the recommendations made by public health officials. COVID-19 was becoming the hot topic of discussion; dominating the headlines on news outlets. I still remember my first time seeing the COVID-19 case-counter posted up on the television for the duration of entire news segments. Since the first time hearing about the virus, the magnitude of COVID-19 started to sink in. Society was about to look a lot different for the foreseeable future, and from that point on, the state of the world started to become more turbulent.

A pandemic is described as a large-scale outbreak of infectious disease that can greatly impact morbidity and mortality over a wide geographic area and cause significant economic, social, and political disruption (Madhav, 2017). COVID-19 is no exception to this. Since cases of this disease were first recorded in the Chinese city of Wuhan in December of 2019 (Wu, Chen, & Chan, 2020), it has spread rapidly throughout the world to infect millions and kill hundreds of thousands (WHO, 2020). Like pandemics of the past, its impact has not been limited to human biology, infection rates, and death counts. The virus has brought a global economic downturn of a magnitude that has not been seen since the Great Depression of the 1930s (United Nations Department of Economic and Social Affairs [UN DESA], 2020) while also inducing fear, worry, and panic from the public. It is also at the forefront of political tensions between China and the United States, with President Donald Trump placing much of the blame on the Chinese government's handling of the outbreak for the current situation. China has also taken aim at the USA for bungling its own COVID-19 response (Peterson, 2020). In some ways, it is quite phenomenal that a virus can have such great consequences, although its effects may also just be a reflection of the vast number of societal connections within the modern world.

Despite the increasing list of negatives that are associated with the pandemic, there have been some silver linings. Tough times have the potential to bring out the best in us. The outstanding dedication exhibited by workers from various industries cannot be understated. From frontline healthcare workers to truck drivers to grocery clerks, many have worked tirelessly to keep our essential systems going through this crisis. Others have sought different ways to help out in the emerging COVID-19 world. To combat the closing of schools and the decrease in access to academic resources, some students have banded together to provide online tutoring. Some have started donation drives to collect personal protective equipment for essential workers. Others have created initiatives to help deliver groceries to those who are more susceptible to the disease. You do not necessarily have to start a worldwide charity to make an impact in others' lives. Simply calling a grandparent to check in on how they are doing can bring some light into their day and show that you are there for them in these unprecedented times.

The increased time spent at home due to social distancing has allowed individuals to reflect and pursue new interests. Many of us lead very busy lives. At times it is rather easy to slip into a state of mindless droning as we go through the motions of our monotonous routines. Some individuals are so goal-oriented that they develop tunnel vision and lose sight of themselves. Regular schedules that juggle careers, academics, family time, and social life do not leave people much time to think about or care for themselves. With non-essential activities being shut down for a long duration, people now have the time to engage in self-discovery. While the world continues to fight against the impact of COVID-19, some have found solace in exploring new things.

Whether it is honing your culinary skills or learning a new instrument, now is a great opportunity to pick up that hobby you have always wanted to pursue.

But while some have utilized the current situation to find new ways to express themselves, others have found it more difficult. Many families have lost loved ones, taking a great emotional toll. The pandemic-induced economic downturn has inevitably caused mass layoffs in the workforce, coercing governments to provide financial support for the public. The implementation of stay-at-home orders have hampered social interactions and may have brought a sense of isolation into some lives. To our readers, I urge you to reach out to your friends and family. Check up on them to let them know they are not alone. Do not hesitate to reach out to others if you are feeling isolated. We live in an age where technology has become an integral part of our society, so make use of it. Send those texts, make phone calls, connect through video calling platforms like Zoom, Skype, and FaceTime. It is a bit different from an in-person interaction, but for now, it will have to suffice.

As the world continues to make technological advances, the field of robotics is sure to become more and more prevalent in our lives. Robotics is a term used to refer to the science and technology of robots: machines that can carry out tasks. Some robots are controlled by humans, while others are automated. These machines are used in a variety of industries and have touched the corners of our world and beyond, scouring the depths of our oceans and reaching for the stars in the cosmos. Autonomous underwater vehicles (AUVs) are unmanned robots that collect data from deep in the oceans. The Canadarm was a remote-controlled robotic arm used in NASA's Space Shuttle Program to move cargo (Canadian Space Agency [CSA], 2018). In parts of Asia, robots have taken up jobs such as cooking in restaurants. Within our homes, some vacuum cleaners are able to memorize the layout of each floor. But the presence of robots in our lives goes far beyond space exploration and futuristic vacuums; the field of robotics has become so integral to the functioning of our society that many common products and services involve the use of these machines. Agriculture, manufacturing of goods, and even healthcare utilizes robots. With social distancing measures in place to prevent too many humans from being in close contact, the importance of robots has increased even further. By the time the next pandemic approaches, robots will likely play an even bigger role due to scientific advancements. Artificial intelligence (AI), which is also called machine intelligence, may be the most compelling concept in robotics. AI robots are fascinating creations in which artificial intelligence allows robots to replicate certain elements of intellectual ability. While its abilities are not at the level of human intellect that is often exhibited by androids in science fiction films, the progress of artificial intelligence in recent years has been astonishing, and it is exciting to imagine what it may be capable of doing in the future. It has already made huge strides in medicine, as a study from 2019 found that the diagnostic performance of deep learning (a form of AI) was equivalent to that of healthcare professionals (X. Liu, 2019). However, this raises a different question; one that not only affects the medicine industry. If artificial intelligence continues to advance, it could revolutionize the workforce and leave many unemployed. Where do we draw the line between employment and automation? The possibilities with robotics are seemingly endless, and it is important to ensure that we reap its benefits to make the world a better place, but also remain aware of its downsides.

Even if it may not feel like it, we are in the midst of a historical event. The events that are currently taking place will be forever etched into our memories and taught in the textbooks of generations to come. People often say that history should serve as a reference point from which we can learn. As much as it sounds like an overused utterance from your high school history teacher, there is truth to it. Viruses are the most abundant entities on our planet (Koonin, 2014). Pandemics, while unwanted, are inevitable. Just over a decade ago, the H1N1 pandemic was the main talking point. A few years before that, SARS was the main concern. We are not even past the COVID-19 surge yet. There is still much to do to save lives and rebuild our systems that have been ravaged. But it is not too early to start learning, especially with the next pandemic possibly looming around the corner. Failures in healthcare systems and policy have not helped the fight against COVID-19, and governments need to start making the necessary adjustments and improvements so that we can be as prepared as possible to handle what comes next. Even if the next big health threat is decades away, working to improve our existing systems will still benefit the public. COVID-19 has raised many questions about how different sectors are run. It is now up to us to address these questions as we move into the future.

Chapter 2

THE CONTEXT OF COVID-19

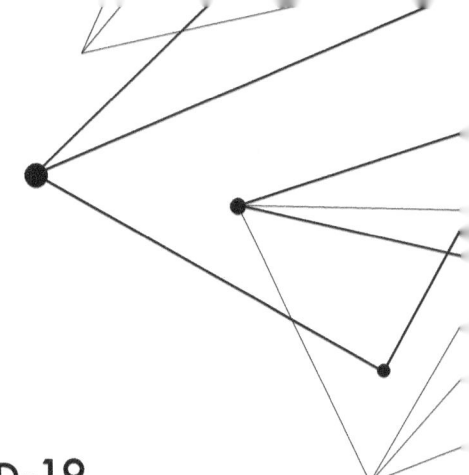

COVID-19, an infectious disease caused by a novel strain of coronavirus, has rapidly taken the world by panic in recent months. Cases of this disease were first reported in December 2019 in Wuhan, a large city of 11 million in central China, when a sudden surge of unusual pneumonia cases overtook the city (Wu et al., 2020). Originating in the Huanan Seafood Wholesale Market, these cases were characterized by fevers, dry coughs, fatigue, and occasional gastrointestinal symptoms, each expressed to varying severity. Although local health authorities immediately took action following the announcement of an epidemiologic alert, the disease had rapidly spread to multiple cities and provinces across China, and soon, to the rest of the world. Canada recorded its first case of COVID-19 on March 9, 2020, when a resident in his 80s at a long-term care facility in Northern Vancouver became infected (Blewett, 2020). By March 11, 2020, the World Health Organization (WHO) officially declared COVID-19 a pandemic.

Coronaviruses are a large family of enveloped, positive, single-stranded RNA viruses that can cause illness in both humans and animals. These viruses are zoonotic, meaning they can transmit between humans and animals. To date, scientists have identified seven human coronaviruses, each classified into the alpha or beta genera based on their genomic sequence. The novel coronavirus, named SARS-CoV-2, falls within the beta genera along with two previously widespread respiratory coronaviruses, SARS-CoV (severe acute respiratory syndrome coronavirus) and MERS-CoV (Middle East respiratory syndrome-related coronavirus; Ciotti et al., 2019). Before 2003, the few known human coronaviruses were only thought to cause mild illnesses, such as the common cold. It wasn't until the subsequent outbreaks of SARS and MERS in 2003 and 2012 respectively that the scientific perspective on the dangers of coronaviruses shifted. Both SARS-CoV and MERS-CoV, as their names suggest, cause severe respiratory syndrome within the infected individual and are often fatal (Wilder-Smith et al., 2020). The SARS outbreak, in particular, was one of the most well-known pandemics in history and also the first documented global epidemic caused by a coronavirus. The mortality rate of SARS was around 10-15%, with its origin tracing back to the Guangdong province in southeast China (Ye et al., 2020). Upon examining the novel SARS-CoV-2 in comparison to the original SARS-CoV, scientists found several similar traits between the two. These traits not only include the expression of common signs and symptoms, but also similarities in the genetic makeup, intermediate host, and disease dynamics. According to a study done by Chan and colleagues, the genome of SARS-CoV-2 shares 82% similarity with that of SARS-CoV (Chan et al., 2020). Together, both viruses also express striking genetic similarity with coronaviruses isolated in bats, suggesting that bats were the intermediate host in the zoonotic transfer of these viruses to humans (Wilder-Smith et al.; Chan et al., 2020). In terms of disease dynamics, the main route of transmission of both viruses is through respiratory droplets, and both target the same protein in the lower respiratory tract for cell entry and viral infection. Although we have seen significant advancement in medicine and technology throughout the years following the 2003 outbreak, scientists have yet to successfully develop a vaccine or treatment for SARS, nor have they developed one recently for SARS-CoV-2. However, through such a comparison, scientists are able to gain a better understanding of this novel coronavirus and extract useful information from previous research to altogether combat the current epidemic.

So we now know what coronaviruses are, but what exactly is COVID-19? To reiterate, COVID-19 is the illness caused by the virus, SARS-CoV-2. According to WHO, the most common symptoms of this disease are fevers, dry coughs, and tiredness. Other less common symptoms include, but are not limited to, aches and pains, nasal congestion, loss of taste, headache, or sore throat. Symptoms usually begin mild and may worsen with disease progression. While close to 80% of those with COVID-19 recover without needing hospital treatment, around 1 in 5 of the infected population become seriously ill and develop difficulty breathing (WHO, 2020). In the worst-case scenario, the severe respiratory syndrome caused by COVID-19 may lead to death. Anyone who experiences a fever, coughs associated with shortness of breath or difficulty breathing, chest pain, and a loss of speech or movement is highly recommended by the WHO to seek immediate medical attention. Older individuals above the age of 60 are more likely to become severely ill if they develop this disease, along with a considerably higher mortality rate than those who are younger. Individuals with underlying conditions such as high blood pressure, heart or lung disease, diabetes, or cancer are also more prone to severe illness when infected (Petersen et al., 2020). In other words, seniors and ill or immunocompromised individuals are two of the most vulnerable populations during this pandemic. However, it is still important to note that anyone can contract COVID-19 and can transmit this disease to those who may be more vulnerable than themselves.

The reason that a regional virus such as SARS-CoV-2 has spread across the globe lies in the degree of efficiency at which it transmits. Initially, SARS-CoV-2 was only believed to transmit from animal to human, with limited transmission among humans themselves. Yet following the rapid increase of COVID-19 cases and the appearance of outbreaks in clusters, data soon confirmed the possibility of person-to-person transmission. As previously mentioned, this disease mainly spreads through droplets from the nose or mouth, typically expelled with coughing, sneezing, or even speaking. Simply by breathing in a droplet expelled from an infected individual, one can contract the virus and develop COVID-19. These expelled droplets may also land on nearby surfaces, such as tables and doorknobs, which we often come in contact with on a daily basis. By touching our eyes, nose, or mouth with contaminated hands, we could contract the virus just the same, only in an indirect manner. This is precisely why global health authorities repeatedly emphasize the importance of hand-washing. Even an action as simple as washing our hands can lower our chances of becoming infected by a significant amount. Another important note is that person-to-person transmission can occur even if the individual is asymptomatic. The incubation period, or the time elapsed between initial exposure to the virus and when signs and symptoms first become apparent, of SARS-CoV-2 can range from anywhere between 2.1 to 11.5 days.12 This means that even without showing the typical symptoms of COVID-19, there is a chance that we have already contracted the virus and can infect other individuals. If anything, this characteristic of COVID-19 simply makes the disease more difficult to contain. With a more comprehensive understanding of the transmission and onset of SARS-CoV-2, we can now see the validity in the prevention guidelines set in place by public health authorities across the globe.

Many changes have been made to our day-to-day lives since the start of this pandemic, with each of us slowly coming to terms with this new "normal". One of the most prominent changes in light of recent events is the implementation of social distancing. By keeping a safe distance of at least 6 feet—or 2 metres—apart from others, this restriction aims to limit the spread of the disease and has proven to be the most effective method in doing so (Government of Canada, 2020). Aside from maintaining a physical distance, social distancing also entails avoiding crowded places or gathering and staying home as much as possible. In other words, distancing ourselves from social interactions. Whether it's in the news, on social media, or in casual conversation, social distancing is typically coupled with other buzzwords such as "self-quarantine" and "self-isolation". Although these two terms are used interchangeably, they each carry a distinct meaning. Self-quarantine refers to the act of limiting interactions between individuals who are not necessarily ill but may have had prior exposure to the virus. It goes hand-in-hand with social distancing and is a precautionary practice that everyone is recommended to follow. Self-isolation, on the other hand, means separating people who are already ill with the symptoms of COVID-19 to prevent them from infecting other people. The standard period of self-isolation is 14 days following the onset of symptoms, as stated by Public Health Ontario. Once the 14 days pass, if the infected individual has recovered, then they are to continue following typical social distancing and self-quarantine restrictions. If the individual remains unwell, then they are to seek medical help immediately.

Knowing that SARS-CoV-2 mainly transmits through respiratory droplets, maintaining physical distance from one another does prove to be an effective prevention method. This, however, brings us to another major addition to our new way of living: mask-wearing. Contrary to the unsupported beliefs and claims endorsed by President Donald Trump, wearing a mask or facial covering does, in fact, limit the spread of COVID-19. For one, masks serve as a physical barrier to prevent the dispersion of respiratory droplets from our mouths, directly obstructing viral transmission. An epidemiologic study published by Health Affairs (Liu & Wehby, 2020) compared the growth rate of COVID-19 before and after the enforcement of mask mandates in 15 states across the U.S. The study found that mandating face mask use in public was associated with a consistent decline in the daily growth rate of COVID-19, slowing the rate by as much as 2.1 percentage-points after 3 weeks. In Canada, mask-wearing largely remains a voluntary recommendation. As of July 2020, the Public Health Agency of Canada (2020) recommends wearing a non-medical mask or face covering at times when physical distancing is not possible, especially in crowded public settings. Otherwise, masks are currently only mandatory for air passengers, in Côte Saint-Luc in Montreal, and in the Wellington-Dufferin-Guelph region in Ontario (Chung, 2020). More than protecting ourselves, wearing a mask allows us to protect others from ourselves. Yet if everyone followed the recommendations and wore masks in public places, then we could all lower our chances of contracting the virus.

Another aspect of consideration which makes COVID-19 a more dangerous contagious disease than the typical seasonal influenza is that scientists have yet to develop a vaccine or a specific treatment for this disease. Once again, this explains why there is such a heavy emphasis on adhering to preventative and cautionary guidelines. In the few months following the emergence of COVID-19, several drugs and medications were proposed to work against the disease. Hydroxychloroquine and Remdesivir are two familiar names that you may have come across in the media. While both drugs exhibit promising effects against SARS-CoV-2 from a molecular and biochemical perspective, the Solidarity trial conducted by WHO (2020) shows that neither drug produces a noticeable reduction in the mortality rates of hospitalized COVID-19 patients. In terms of vaccines, although clinical trials are taking place all around the world, it can take several years before a new vaccine is fully developed. Although research for treatment and immunity is still underway, well-developed detection methods for the virus do exist. Two kinds of tests are available for COVID-19: a viral test which indicates a current infection and an antibody test which reveals previous infections. Anyone who exhibits symptoms of COVID-19 is able to request for a test from their healthcare provider or health department, who then decides to administer the test.

The emergence of this global epidemic has thrown every one of us into a state of uncertainty. For researchers across the globe, COVID-19 poses a new challenge and opens the door for further investigation on infectious disease. For healthcare workers, they have been dragged into yet another battlefield to fight against a novel disease. Although the current situation in regards to COVID-19 stands as is, we cannot predict what the situation will be like in the future; whether it is a week, a month, or even a year from now. Regardless of your position or status, this period of adversity has likely affected you in some way. The best course of action for us as individuals is to follow the public health guidelines and to play our part in putting an end to this pandemic.

Chapter 3

THE BURDEN ON HUMANS

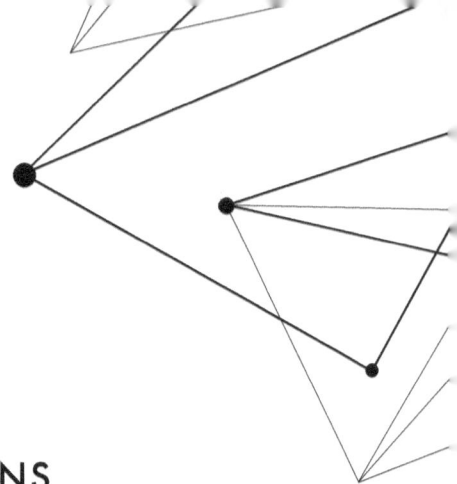

COVID-19 has the potential to elicit a plethora of physiological responses should it be contracted, wreaking havoc on the human body. However, it is estimated that 70% of patients are asymptomatic or display very mild symptoms (Cascella et al., 2020). In the remaining 30% of patients, there is a respiratory syndrome until severe respiratory failure is reached, which may require admission to ICU. Mild cases often present with symptoms of an upper respiratory tract viral infection, which includes mild fever, cough, nasal congestion, headache, vomiting, diarrhea, and loss of taste and/or smell. More severe cases may exhibit severe pneumonia and acute respiratory distress syndrome (ARDS), which requires most patients use a mechanical ventilator that helps them breathe. While many are aware of COVID-19's burden on the human body and its subsequent physiological systems, it also manages to exert its influence on some of the larger, branching systems throughout the current world.

As of July 13th, 2020, there have been 108,155 total cases of COVID-19 and 8,790 COVID-related deaths in Canada (Canada, 2020). Across the entire globe, there have been 12,964,809 confirmed Coronavirus cases and 570,288 confirmed deaths (WHO, 2020). Needless to say, COVID-19 has had a profound impact on populations - one that has lived up to its label of a global pandemic that was announced by the World Health Organization on March 11th, 2020. This global outbreak caused many countries to enforce various measures to help reduce the spread of the disease. The concept of "flattening the curve" is often referenced and advocated for by governments and health officials. It refers to the combination of strategies to slow the spread of COVID-19, thus preventing hospital capacities from being exceeded (Kenyon, 2020). Some countries such as Taiwan and South Korea have been applauded for managing to do this successfully through well-executed public health responses (Cavallo, Donoho, & Forman, 2020). However, other nations have struggled. Italy's National Healthcare Service (NHS) faced immense pressure from the massive influx of COVID-19 patients, forcing the nation to reflect upon the impact of the decentralization and fragmentation of health services over the past decade—one that included financial cuts of over €37 billion to the NHS (Armocida, Formenti, Ussai, Palestra, & Missoni, 2020). The United States' decentralized healthcare system has drawn harsh criticism from media outlets and citizens alike, especially when comparing their handling of the pandemic to a country with similar geographical and cultural characteristics such as Canada. American advocate for health insurance payment reform and New York Times bestselling author Wendell Potter took to social media and tweeted, "You learn a lot about a health-care system when a global crisis hits & different nations have different results. Canada's single-payer system is saving lives. The U.S. profit-driven model is failing" (Ho, 2020).

Despite the praise from some people south of the border, COVID-19 has elucidated some of the problems existing in Canada's healthcare system. It is estimated that between 62% and 82% of COVID-19-related deaths in Canada occurred among residents of long-term care facilities. In Ontario, 5 of the 7 health care workers who have passed away from COVID-19 were personal support workers in long-term care homes (Holroyd-Leduc & Laupacis, 2020). While the staggering impact of COVID-19 among the elderly population is unsurprising due to their age and associated physiological factors, many have pointed out that the same amount of effort put into the hospital sector should have also been put into long-term care. Measures such as aggressive testing and ensuring that personal support workers are to only work in one facility were only implemented after multiple outbreaks had occurred. This, in combination with a lack of personal protective equipment for these facilities, highlights the undervaluing of this healthcare sector, where some support workers earn as low as $14 per hour (Holroyd-Leduc & Laupacis, 2020).

While there is a massive focus on the impact of COVID-19 on our healthcare system, it has also ravaged both the local and global economies. The mandating of temporary closures for non-essential services and businesses combined with the recommendation of social distancing has left many store owners and company executives reeling with a massive decrease in revenue. Even restaurants that have somewhat adapted to the technological trends of UberEats and other food delivery services have struggled because of their COVID-19-related inability to seat customers. Even with social gathering restrictions slowly lifting, the requirement for groups to remain socially distanced is likely to stick around, thus reducing the number of people allowed to be physically present in stores.

The impact of COVID-19 does not just stop at local businesses. The Dow Jones closed down -23.2% for the first quarter of 2020, marking its worst first quarter in history as many investors engaged in panic selling (Imbert, Stevens, & Fitzgerald, 2020). Global economic output expects to lose $8.5 trillion over the next two years, making this the sharpest economic contraction since the Great Depression that took place in the 1930s (UN DESA, 2020). Canada alone saw the loss of over 1 million jobs in March of 2020, which is the largest single-month decrease since records began in the 1970s (BBC, 2020). Around the world, the pandemic is also expected to cause 34.3 million people to fall below the extreme poverty line by the end of 2020, with an additional 130 million possibly added to that category by 2030 (UN DESA, 2020). Should these forecasts come to fruition, it will only magnify the socioeconomic disparity within our society. Despite the efforts of various governments to provide financial aid to those in need, COVID-19 has disproportionately affected low-skilled, low-wage workers in comparison to more highly-skilled counterparts. Those with lower socioeconomic status are (1) more likely to live in overcrowded areas and (2) tend to be employed in occupations that do not provide opportunities to work from home; both make them more susceptible to contract COVID-19 due to a reduced possibility of social distancing (Patel et al., 2020). People in these groups are more likely to have unstable working conditions and incomes and are therefore greatly impacted by the sharp economic downturn that the pandemic has induced. The characteristic of a generally lower level of education and skill also tends to place them at the top of the firing list when company executives need to downsize. Aside from the general economic downsides that COVID-19 has set ablaze, there are astronomical financial implications for some who are unfortunate enough to contract the virus in a country with a privatized healthcare system. In one heavily publicized case, a 70-year-old American survivor of COVID-19 incurred a hospital bill of $1.1 million for services such as intensive care, room sterilization, and use of a ventilator (CTV News, 2020). Fortunately, the costs were covered by a government insurance program for the elderly called Medicare, but this is not the case for many others.

Long-term systemic health and socioeconomic inequities place an increased burden on certain ethnic minority groups, especially in these current times. As of June 12, 2020, indigenous and black persons face COVID-related hospitalization rates five times greater than that of caucasian persons in the US (Centers for Disease Control and Prevention [CDC], 2020). For the Hispanic community, persons are four times more likely to be hospitalized due to COVID-19 compared to non-Hispanic white persons (CDC, 2020). The fact that the pandemic has disproportionately affected these minority groups is caused by a variety of factors. Many members of these communities are more likely to live in densely populated areas, which exemplifies in reservation homes as a result of residential housing segregation. Those living in these potentially overcrowded areas may find it harder to practice social distancing, thus propagating the spread of the disease. Also, members of ethnic minority groups are more likely to have lower levels of education and lower-paying jobs (CDC, 2020). As previously mentioned, these socioeconomic factors often result in occupational environments that increase the risk of contracting COVID-19.

One racialized minority group, in particular, has been greatly affected by COVID-19, but not in a genetic or epidemiological way. Since the global outbreak began, countries such as the United States and Canada have seen an increase in anti-Asian behaviour. Whether this is strictly a response to the relation between China and COVID-19 or because the pandemic has enabled hidden racism to manifest remains unclear, although it may be an amalgamation of both. Nevertheless, it is unacceptable. As of the 22nd of May, Vancouver saw 29 cases of anti-Asian hate crimes within the year 2020, a massive jump for a city that saw just 4 cases the previous year (Judd, 2020). In America, the Asian Pacific Policy & Planning Council documented over 1000 reports of hate crimes directed towards Asians between March 19th and April 1st of 2020 (Chen, Trinh, & Yang, 2020). This behaviour trend reflects in social media, with Twitter seeing a surge in Sinophobic slurs beginning in late January 2020 (Chen et al., 2020). The anti-Asian sentiment has not been helped along by United States President Donald Trump, who has faced criticism for exacerbating the association between the pandemic and China by using phrases such as "Chinese virus" and "Kung Flu" to refer to COVID-19 (B. Lee, 2020).

The widespread reach of the COVID-19 pandemic into many facets of our lives also raises questions regarding the psychological impacts. Since the turn of the century, suicide rates are significantly increasing in the United States. Data from 2018—the latest data available—shows the highest age-adjusted suicide rate the country has seen since 1941 (Reger, Stanley, & Joiner, 2020). The social distancing measures implemented to combat the spread of the virus may have some adverse effects regarding suicide and mental health. The sudden economic downturn, which includes mass layoffs in the workforce and historic drops in the stock market, may factor into increasing suicide rates. Recessions are often associated with higher suicide rates in comparison with times of relative prosperity. Leading suicide theories also note the role that social interactions play in suicide prevention. Individuals contemplating suicide may lack a strong social network compared to those at lesser risk, and suicidal thoughts are associated with social isolation. Thus, the notion of social distancing does not seem like such a great idea from a prevention standpoint, as suicide rates are expected to increase in the face of the pandemic. However, since it is the status quo for "flattening the curve," it is vital that individuals find alternatives to connect with others. In an age where technology is extremely prevalent, many have taken advantage of its benefits by shifting in-person events into "virtual" gatherings. Video calling applications such as Zoom, WebEx, and Skype have become important tools in workplaces and educational institutions, allowing users to speak with each other "face-to-face" in real-time. Even those looking to practice their religion have started to make use of these services to attend events like weekly mass.

As expected with a global pandemic, COVID-19 has inflicted the kind of burden that rarely occurs throughout history. It has taken hundreds of thousands of lives and greatly altered even more. It has placed an enormous amount of pressure on various systems intertwined within our society, which consequently impacts the very populations that rely on them. But despite its profound impact on humans, it has also managed to unravel and magnify some of the issues with these systems. Socioeconomic disparities and negligence towards specific sectors are exposed to the public eye due to the disproportionate impact that the pandemic has had on affected groups. Controversial topics such as socialized healthcare systems are now gaining more attention and possibly more traction due to some countries' experiences in dealing with the large-scale effects of COVID-19. This surely will not be the last pandemic to emerge, and the world needs to learn from these lessons moving forward so that we can be prepared to deal with whatever comes next. However, the focus should not just be on improving the response to pandemics; there needs to be a focus on improving the systems that have failed different groups, even without the presence of a coronavirus. As humanity starts to emerge from the COVID-19 shadow, it needs to make sure to address some of the self-imposed burdens that were existent long before the virus reached patient zero.

Chapter 4

THE ADVANTAGES OF ROBOTICS IN PANDEMIC SITUATIONS

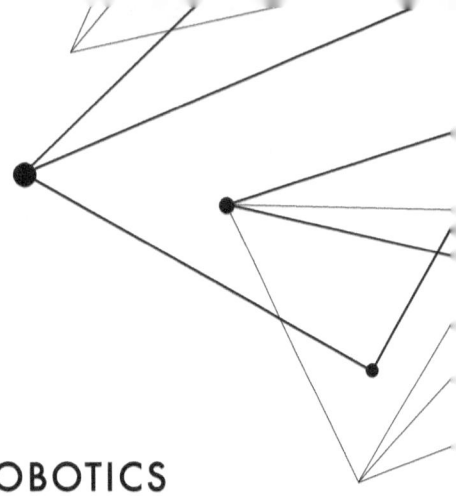

In our search for new and innovative strategies to alleviate the immense burden which COVID-19 has placed on humans, a spotlight has gathered upon the emerging notion of automation. With the rapid advancement of modern-day technology, robots and artificial intelligence are no longer theoretical concepts actualized in science fiction, but real inventions employed in everyday life. Although people have often raised concerns over job losses and privacy issues with regards to the incorporation of robots into society, amid a global epidemic, these risks are largely outweighed by the potential benefits which robots can bring. During the 2015 Ebola outbreak, robots made a significant contribution to three broad areas of disease management: clinical care, logistics, and surveillance (Yang et al., 2020). In terms of clinical care, robotic use was engaged in various ways throughout each stage of the process, from disease prevention, to screening and diagnosis, to patient care. Within logistics, robots also played an active role in the handling and transport of contaminated waste. By synthesizing the data collected during this process, these robots altogether allow us to closely monitor the progression of the outbreak, thus contributing to surveillance. In each of these areas, robots proved to be an advantageous addition to the healthcare system. Each of these advantages we will now further examine as they apply to the present-day COVID-19 pandemic.

Given the highly contagious nature of COVID-19, the most effective means of preventing the spread of infection is to limit human-to-human interaction. This is where the advantages of adopting modern-day technology such as robots and telemedicine come to light. Whether robots are programmed to perform tasks entirely on their own or through manual control, the application of these machines considerably reduces the need for human personnel, which consequently reduces person-to-person contact. When it comes to major outbreaks such as the COVID-19 pandemic, the area of clinical care in which robots prove to be the most beneficial is screening and testing. As mentioned in chapter 2, two types of tests are currently available for the screening of COVID-19: viral tests and antibody tests. Although antibody tests, which screens for past infections, have been shown to produce false-negative results, both types of diagnostic can quite accurately identify the presence of an infection. Regardless of the type of test conducted, both testing processes involve collecting nasopharyngeal or oropharyngeal specimens—or in simpler terms, nose and mouth swabs. The act of swabbing directly exposes the person conducting the test to the respiratory droplets of the patient, putting them at a higher risk of contracting the virus. As swabbing is a relatively mechanical action, robots can instead be employed to collect samples from patients in an automated process. Not only would this reduce the chance of infection of the clinical staff, but it would also free them up to complete other, more technically complex, tasks required for the testing process.

Although we have addressed sample collection, this procedure is then followed by handling, transfer, and laboratory testing, all of which further exposes clinical staff to the virus. Luckily, the use of laboratory robots for the transport and delivery of lab samples is already prevalent in several healthcare facilities across the world. In fact, you may have already encountered one at your local hospital. During an infectious outbreak, these robots can be operated in a similar manner to deliver medications to infected patients in quarantine, which once again reduces risks of transmission by allowing hospital staff to maintain distance. Even laboratory testing, the final step of the diagnostic process, can be automated. Given the proficiency of artificial intelligence nowadays, automated real-time assays can be integrated with robot-administered swabbing to allow for rapid in-vivo detection of pathogens. A mobile robot equipped with a thermal sensor, programmed to use swabs, and automated to conduct real-time assays can easily execute a process that typically requires an entire team of healthcare workers. The potential for robots to complete multiple tasks simultaneously, therefore, demonstrates the immense versatility of robotics and AI as a whole, and reveals to us the various improvements it brings to clinical care.

Another notable challenge of containing an outbreak of this extent is the inability to conduct tests at a larger scale. A shortage of available COVID-19 test kits has been a recurring issue in multiple countries across the world, mainly due to the diminishing supply of global resources as a result of increased demand. This is an especially prominent issue in the U.S., as the rapid surge in cases repeatedly exceeded the country's testing capacity despite the constant production of more testing kits. In Canada, COVID-19 tests are prioritized for those with serious symptoms, pre-existing conditions, and healthcare workers, while those with mild symptoms are encouraged to self-isolate. However, David Bowdish, Canadian Research Chair in aging and immunity at McMaster University has stated that by restricting testing to more serious cases, it has led to a wider community spread of the disease (Cousins, 2020). With these challenges, scientists once again look to robots and artificial intelligence in finding an alternative solution to the problem.

As of May 2020, Imran and colleagues developed and tested an AI-powered screening solution for COVID-19 that is deplorable via a smartphone app named AI4COVID. This app is designed to closely analyze the cough sounds of the user and produce a corresponding diagnosis within 2 minutes (Imran et al., 2020). Although the technicalities and accuracy of this app are still being refined, the results of the initial trial indicated a promising outlook for the potential incorporation of AI in large-scale screening. Aside from cough detection in COVID apps, scientists are looking to use mobile robots for temperature measurement in public areas, by equipping them with thermal sensors and vision algorithms. Automated camera systems can also be installed to screen multiple individuals simultaneously in crowded areas. By setting up automated sensing systems at ports of entry in high-traffic, high-risk areas such as hospitals or clinics, it would increase both the efficiency and coverage of screening. Another major benefit of using AI or computerized systems for screening is the ability to establish a shared database with the collected data, all of which are readily accessible. In hospitals or clinical settings, this data could also be linked to the health database to directly update patient information with screening results. Therefore, not only can robots improve the screening process itself, but they can also help us gain a better understanding of the disease progression as a whole.

A brief glimpse at the current healthcare situation would reveal that hospitals all over the world are oversaturated as a result of the COVID-19 outbreak. In addition to the existing hospitalized patients, hospitals now have to accommodate those who are severely infected with COVID-19 while also practicing physical distancing. As hospitals scrambled to prepare for the surge in COVID-19 cases, several non-COVID related services had to be put on pause, including non-emergent surgeries, clinical checkups, and treatments. Although services are gradually resuming with the stabilization of COVID cases, hospitals remain high-risk environments for both patients and visitors. This brings us to the topic of robotic surgery, which has skyrocketed in popularity in recent years due to its numerous clinical advantages, from increasing efficiency to minimizing surgical invasion. In terms of its advantages in an infectious environment, the minimally invasive quality of robotics surgeries reduces the amount of open area exposed to surgical staff, effectively limiting viral transmission via blood or surgical plumes. A minimally invasive procedure also correlates with a shorter recovery time post-operation, which consequently reduces the amount of time that patients would have to spend in a hospital (refer to Chapter 14 for more on robot-assisted surgery). Clinical consults are also migrating to teleoperation and other robot-associated communication methods in replacement of in-person visits to minimize exposure for patients. With that said, the implementation of innovative technology in healthcare will overall serve to improve the quality of care which patients receive during these unprecedented times.

In addition to patient care, robots can provide a helping hand with other more logistical tasks in clinical settings to reduce manual labour and relieve the burdens of frontline workers. One example is implementing robot-controlled noncontact disinfection of contaminated surfaces using ultraviolet (UV) light devices (Yang et al., 2020). The World Health Organization (WHO, 2020) stated that coronaviruses can survive on inanimate surfaces from anywhere between a few hours to several days, making it all the more important for proper disinfection procedures to be practiced in public places. UV light is effective in reducing contamination on high-touch surfaces in hospitals (Yang et al., 2020). Therefore, when combined with the autodetection feature programmed in robots, this system would allow for the disinfection of contaminated spaces at a larger scale, in a timely and effective manner. An important note to keep in mind is that there are many highly contagious environments during a pandemic that are not suitable for human workers but can be tolerated by robots. In the end, robots are machines, with a much higher work capacity than their human counterparts and are capable of tirelessly performing a variety of tasks that we are unable to. All of which come together to showcase the potential of robots supporting healthcare providers.

Now that we have closely examined the advantages of robots in pandemic situations from a clinical and healthcare perspective, it's time for us to take a look at the role of robots through a different lens. In July 2020, Othelia Lee and Boyd Davis (2020) of North Carolina University proposed a pilot study of adapting "Sunshine"—a socially assistive chat robot for older adults with cognitive impairment. The study was conducted with the hopes of developing an innovative method to provide company to older adults with cognitive impairments in long-term care facilities, who are already distanced from social life during this COVID-19 pandemic. This pilot study exemplifies the versatility of robots, by demonstrating the social competence of artificial intelligence. While humans are equally as capable, if not more, of conversing with the older population, the presence of human interaction immediately puts the older adults in a vulnerable position for contracting the virus. In the context of a pandemic, the socially assistive robots would serve as the optimal solution, as it not only provides contactless social support but also keeps the individuals safe and COVID-free.

In a technological regard, COVID-19 has certainly hit us at an awkward time. Although we have developed advanced AI systems and are beginning to explore the practicality of robots in healthcare, we have yet to create robots advanced enough to fully assume the duties of healthcare workers. However, there are still several advantages and benefits which our current technology can provide in a pandemic situation. By deploying robots to complete tasks in clinical settings, we can reduce exposure and limit viral transmission, reduce manual labour, and free up clinical staff. Advancements in AI have led to the operation of telecommunication and telemedicine, both of which help to decrease clinical traffic. The ability of automated systems to collect data on a larger-scale provides us with useful insight into the epidemiology of the disease, thus assisting us to contain the outbreak. Additionally, robots can also play a beneficial role in personal and long-term care homes to improve the quality of life of the quarantined seniors. Regardless of the extent to which robots can help, the possibilities and advantages listed above shine a light on the immense potential that robots hold for the healthcare system as they join the frontlines of the pandemic.

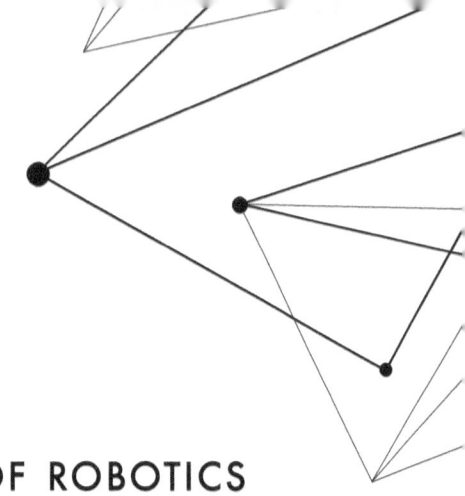

Chapter 5

THE DISADVANTAGES OF ROBOTICS IN PANDEMIC SITUATIONS

Despite the growing demand for robotics and automation in healthcare, a number of limitations still hold us back from fully transforming into a robot-run society, especially during the current pandemic. One of the major disadvantages of using robots stems from their technological shortcomings. Although we have made significant advancements in perfecting artificial intelligence (AI), the majority of present-day robots still exist as programmed machines that function in response to external commands. In order to smoothly integrate robots into the workplace and ensure that they improve—rather than hinder—the workplace efficiency, the surrounding work environment must be controlled. Automated robots are especially prone to environmental discrepancies, as even a small, unexpected movement could disrupt their workflow and result in malfunction. With this in mind, most of the current generation of robots are only assigned to complete simple and repetitive tasks to minimize the margin for error. In retail, robots are used limitedly to track inventory, clean spillage, and monitor cash registers. In administration, robots assist with data entry and other mundane managerial tasks. In healthcare, robots are mainly used for sample delivery, with more sophisticated robots slowly working their way into the operating room to assist with surgical procedures. While robots can certainly complete these routine tasks to perfection, it is difficult to say whether they can do the same in an unpredictable and highly variable environment, or in a pandemic.

In light of the COVID-19 pandemic, hospitals across the globe are faced with sudden increases in demand for clinical care and have become oversaturated with infected patients. This endless influx of patients continuously introduces new variables to the workplace, which only serves to hinder the ability of medical robots to perform their job. Even the mere presence of more people in the hospital is an obstacle in itself, as it becomes more difficult for automated robots to navigate the hallways without colliding into any staff or patients. A similar disadvantage is seen with using robots as we transition to a post-pandemic world. Following the initial outbreak in Wuhan, AI-controlled driverless vehicles were deployed in the streets for delivery to minimize human interactions during lock-down (Arthur & Ruan, 2020). However, although this was a perfect solution for transporting resources at the time, these driverless vehicles would likely experience difficulty when navigating the streets through everyday traffic, which is bound to resume once the city reopens. This is not to say that robots cannot be operated once the environment has been brought back under control. In a hospital setting, hallways can be sectioned off to provide designated lanes for lab delivery robots, which would effectively control the flow of traffic and allow for easy travel. If cities also did the same and introduced designated lanes for driverless vehicles in densely populated areas, then the application of automated robots for delivery could still be maintained in a post-pandemic setting. As easy as these modifications seem, it would still take time to implement them and even more time for people to adapt to them, bringing us to yet another disadvantage of using robotics in pandemic situations: an insufficient time frame.

A shortened time frame is perhaps the most prominent limitation that holds us back from deploying robots at once to combat this pandemic. The AI-regulated functions expected from robots for them to assist in a pandemic are very specific to the situation, which subsequently requires specific robotic programming. As mentioned previously in chapter 4, this would include programming precise movements to perform COVID testing, implementing thermal sensors for large-scale screening, and potentially incorporating real-time assays to speed up the diagnostic process. While these proposed robotic additions sound promising, each of them still takes time to develop, which we can hardly spare, considering the fast-moving dynamic of the current situation. In fact, researchers have revealed that many of the capabilities required for robotics at this time have not yet been developed or properly funded (Strickland, 2020). So unless scientists can develop and manufacture fully functional, pandemic-ready robots in the next few months, the best course of action would be to take the time now to prepare them fully for any other uncertainties that may arise in the future.

Not only would it take time to develop functional robots for COVID frontlines, it would take more time before these robots can fully integrate into the workforce. For proper and smooth integration, two components must be considered: first, delivering formal training for robotic use; and second, getting comfortable with human-robot interactions. Just as with any other new machines introduced to the workplace, employees would have to receive mandatory training in order to work safely with robots. In a pandemic situation, dealing with a contagious disease already puts a strain on the maintenance of workplace safety, yet the addition of robots merely introduces another variable that now needs to be considered. Healthcare professionals need to learn how to work alongside robots and become knowledgeable about the precautions that need to be taken to help facilitate interactions between robots and patients. In general, the incorporation of medical robots could very likely disrupt the existing workflow of the healthcare system. Although this challenge can be overcome by adopting a gradual approach when introducing robots to the medical front, the process must be accelerated in order to accommodate the urgent demands of a pandemic.

Getting trained to work with robots is one thing, but getting comfortable to work with them is another. Ever since the release of The Terminator and subsequent robots-take-over-the-world movies, there has been a lot of prejudice against robots and AI in the real world. In a healthcare setting, even if medical professionals are able to work together with their mechanical counterparts, the patients may not be as accepting. Guang-Zhong Yang, founding editor of Science Robotics and dean of the Institute of Medical Robotics at Shanghai Jiao Tong University stated in a CNN article (Strickland, 2020) that "in medical robots, patient safety and effectiveness are highly-regulated areas. But people also need a comfort level with them in order to feel safe—especially if one is swabbing inside their nostrils." The point of using robots in pandemic situations is to replace human-human interactions with human-robot interactions where possible in order to limit viral transmission.

While all this effort would be aimless if the humans in question refuse to interact with robots at all, we cannot blame anyone for feeling unsafe. This challenge of gaining trust from patients is further augmented by the stressful climate of the pandemic. Whether the patient seeking clinical care has tested positive or negative for COVID, they are equally as vulnerable and would likely feel neglected if they were tended to by a robot than a real human being. According to an article published by Forbes (Jain, 2020), the Circolo di Varese hospital in Italy has recently adopted a team of robotic nurses to assist with the hospital treatment of COVID-19 patients. This team of robots contributes to clinical care by constantly monitoring ICU ventilators, measuring oxygen levels and blood pressure, and allowing staff to communicate with patients from a distance. Dr. Francesco Dentali, the hospital's ICU director, mentioned that it is critical to explain the aim of the robots to the patients as the first reaction to the robotic team is usually not positive, especially for older patients. Although we can encourage acceptance by educating patients on the benefits of robotics, there will always be people who feel uncomfortable interacting with machines.

On the topic of identifying limitations with regards to human-robot interactions, we have to realize that the current generation of robots are still socially inept, and cannot express the same compassion, empathy, and other emotions as humans can. The emotional connections which humans can provide for one another are extremely important during uncertain times such as the current. Only humans could be able to offer comfort and provide relief for the psychological stress and anxiety brought upon by the pandemic. Again, looking through a healthcare lens, many patients who visit the hospital for clinical treatment unconsciously expect a certain degree of assurance and emotional support for their predicament. A similar lack of connection is expected from social robots that are created to provide company to individuals in quarantine. Regardless of how advanced modern AI may be, it still falls short to humans when it comes to showing empathy. Therefore, the inability of robots to comprehend complex emotions makes it challenging to use them on a more interpersonal basis.

Another common concern surrounding robotics is the issue regarding privacy and data protection when it comes to machine learning. The use of robots in healthcare immediately brings to mind the existence of patient confidentiality policies and whether or not data collection through AI should be considered as a breach of policy. One of the main proposals for using robotics during this pandemic is to compile COVID screening results on a shared database, which allows public health authorities to monitor the outbreak. Although this is a convenient and innovative method of tracking the spread of disease from person to person, it does raise some ethical concerns with regards to privacy. Privacy has been a popular topic for debate when it comes to evaluating the ethics of robotics and AI. It comes to a point where the benefits of AI technology must be weighed against its risks, and experts must carefully consider how this technology should be integrated into clinical practice to minimize these risks.

The final disadvantage of robotics is the fear that robots are taking over our jobs and inflating unemployment rates. As most of us know, the COVID-19 pandemic not only impacted our healthcare systems but it also heavily impacted the global economy. From the initial outbreak in early January until now, many economically advanced countries across the world witnessed an abrupt spike in unemployment rates as businesses are forced to shut down and workers to return home. The North American job market has never looked as bleak as it did in April 2020, when the coronavirus had reached its peak. From February to April, Statistics Canada (2020) reported a total loss of 3 million jobs across the country, with COVID-related absences from work amounting to 2.5 million. Unemployment rates in Canada also hit a record high of 13.7% during the following month in May 2020. Compared to Canada, this COVID-induced economic collapse hit the United States even harder as the number of unemployed persons increased by 15.9 million from February to April, with an unemployment rate of 14.7% at the time (U.S. Bureau of Labor Statistics, 2020). With all statistics considered, we can see the detrimental effects which robotics can pose on the current job market. In compliance with social distancing, the first jobs that are at risk for automation are those involving high-touch environments, such as retail, restaurants, and recreation. A study conducted by Oxford University in 2017 evaluated the impact of robotics on the job market by calculating the share of jobs that can be automated within the next 15 years. Regarding the high-touch, high-exposure industries mentioned previously, this study predicted that by 2035, technology should be able to automate 86% of restaurant jobs, 76% of retail jobs, and 59% of recreation jobs (Frey & Osborne, 2017). If we truly shift towards automation in the post-pandemic period, when unemployed workers are just starting to search for new employment, the additional robot-related unemployment would only add to the problem.

After evaluating both the advantages and disadvantages of robotics in pandemic situations, should we or should we not rely on robotics to help us make our way out of this predicament? The bottom line is: we have to weigh both sides to determine the best course of action that would ultimately bring about more benefits than harm. Although valid, the limitations and challenges of robotics mentioned above can all be resolved or worked around to a certain degree. With the issue of an unpredictable environment caused by the pandemic, we can start making the necessary modifications now so that robotics can be readily operated when the next pandemic hits. By starting the preparations early, it also eliminates the limitation of an insufficient time frame as we would not have to scramble to find innovative solutions a second time. As for the job market, we can use this pandemic as an opportunity to expand employment to robotics and technologically-related fields, potentially creating new job openings. However, privacy remains an ethical dilemma of artificial intelligence, and will always be a topic for heated debates. On a final note, this is not the first time that humans looked towards robots for reinforcement in a crisis. Similar plans for robot assistance were developed following the 2015 Ebola outbreak, yet the plans never took off, due to the gradual loss of motivation and funding over time. With this lesson from the past, we should aim to achieve a sustainable approach to robotics research and development, so that we are prepared for future uncertainties as individuals of the 21st century, the information age, should be.

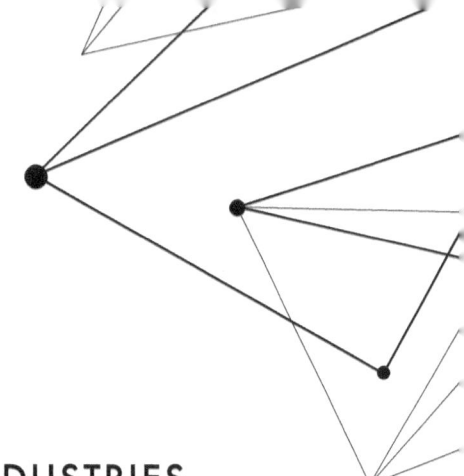

Chapter 6

ROBOTICS IN RETAIL INDUSTRIES

A rumbling stomach. A sign of hunger. You pull yourself out of the task at hand and walk downstairs to your kitchen. There is enough food in there for a couple of meals, but the supply is running low and it is time to stock up. After you fix yourself a quick bite, you grab your car keys and drive to the local supermarket. Somehow, you manage to find everything you need in that labyrinth of aisles, albeit taking a bit longer than you expected. After all, you did have quite an extensive grocery list. As you pay, the cashier kindly greets you with a smile and wishes you along in your day.

Once in the parking lot, you remember that tomorrow is your first day at a new job. Always being one to look good (you are the epitome of the phrase 'dress to impress'), you then decide to head to the mall to pick up some new clothes. One of the staff in the clothing store helps you pick some nice-looking fits and colour schemes. Happy with the vibe the outfits give off, you bring your new clothes to the front and hand over the appropriate amount of cash to the worker behind the desk. Now it is time to explore other parts of the mall; stores that are much more than the enclosed area with products and price tags. There is much involved in this sector; those items lining each wall needed to be manufactured, packaged and shipped to stores that customers can visit to spend their money. Each item, big or small, is part of a massive industry with many resources, processes, and people involved.

These are just some examples that highlight the prevalence of the retail industry in our everyday lives. Even when we do not want to leave the house or are too busy to stop by a store, retailers with an online presence such as Amazon can ship products directly to us. Technology has become much more advanced over time, with industries making shifts towards automation being representative of such change. Robotics, AI, and machine learning enable retailers to perform processes with greater productivity, efficacy, and efficiency, so it is only a matter of time before applying these concepts to the retail industry. In fact, this has already become an integral part of many companies. For instance, big-box retailers such as Staples, Gap, The Office Depot, and Amazon make use of robots in their distribution centres (Joshi & Kumar, 2014). These machines take merchandise and bring it to workers to be packaged and shipped. Implementing this autonomous technology has had its share of quantifiable benefits; they allowed Amazon to ship packages in 13 minutes from pick stations, compared to the average 1.5 hours at older centres. Subsequently, its usage within the company has been widespread. Amazon has 15,000 robots across 10 warehouses in the United States to help mobilize packages (Joshi & Kumar, 2014). Companies are also looking to take advantage of drone technology, and Amazon has publicly discussed plans for drone delivery in the future (Grewal, Roggeveen, & Nordfält, 2017). Even customers in the retail industry have hopped onto the technological trend, with 62% of users of voice-based digital assistants such as Siri or Alexa plan to purchase something through these smart devices within the next month (Belanche, Casaló, Flavián, & Schepers, 2019).

The specific application of robots in various retail functions has its benefits. When used in conjunction with artificial intelligence, they can be of great help in customer assistance. Robotic shopping carts can assist customers by guiding them to the right aisle and perform or cancel checkouts when items are either added or removed. This is a massive time-saver for shoppers, who will not have to spend time racing across stores to get to the correct aisle. Additionally, these machines are also able to provide a personalized shopping experience. They can infer preferences from facial expressions and body language and then recommend items based on this information combined with your budget and previous purchasing history. Robots can also be applied in less complex contexts by performing useful and convenient services such as currency exchange, price comparisons, or gift-wrapping.

Additionally, robots can support employees and managers. They can particularly play a role in accelerating the fulfillment of tasks such as carrying out pick-up items from online orders. When an item is ready for pick-up, a robot can move to the store inventory, identify the item ready for pick-up, and deliver it at the pick-up counter. The intelligence of these robots also allows them to perform routine checks and audits. At least 25% of items labelled 'out of stock' are actually in the store but are not on the correct shelves (Joshi & Kumar, 2014). Robots can monitor inventory and then use their motor skills to replenish depleted shelves or return items to their correct place. The surveillance capabilities of robots can also help mitigate shrinkage, which occurs when there are fewer items in stock than in the inventory list due to damages, errors, and theft. US retailers lose an average of $54 billion in sales per year due to shrinkage (Joshi & Kumar, 2014). Implementing robots can save companies vast amounts of money, ensure that products are available and ready for consumers, and possibly deter thieves due to their surveillance.

It is impossible to implement robots in the retail industry without experiencing any disadvantages. Detractors may note dehumanization and social deprivation as the downfalls of this technology. The friendly greetings from that supermarket cashier? A thing of the past. The kind person who helped you pick out an outfit has been replaced by a robot who does the same thing. Even though these robots can out-perform their human counterparts in these tasks, the removal of the human essence and interaction from the retail experience is something we should be wary of, particularly in light of the current COVID-19 situation. The psychological impacts that COVID-19 has had on humans as a result of reduced social interaction have been worrying for the public, despite distancing measures only being active for a relatively short time frame. Just imagine if the retail industry became fully automated. This would not be an ordeal that merely lasts a few months; the absence of human interaction in various commercial contexts would be the new normal.

Another point that may surface when discussing the use of robotics in an industry is the adverse effects it will have on employment rates. Automation can impact employee morale, who may see the machines as competition for jobs. Robots can provide homogenous output, have virtually endless memory and access, are easily upgradable system-wide, can engage in pattern recognition, and can take on unattractive jobs (Wirtz et al., 2018). They are better than us in virtually every relevant category when it comes to performing tasks. The wise economical choice would be for retail companies to begin further integrating robotics into the workplace. Many players in the industry already rely on robots for a multitude of tasks, so it is likely that they may just look to fully automate processes, thus maximizing profits. This would put an astronomical number of workers into unemployment. Think about the hands that are potentially involved in making and selling something as simple as a t-shirt: it had to be stitched, coloured using dye, put out to dry, packaged neatly, shipped by some method of transportation, unboxed in the inventory room, put out onto the clothing rack for display, and sold and bagged at the front desk. All that for a single t-shirt, and yet many lives that could be changed if they were to be replaced by autonomous robots.

Robotics and AI also present challenges and questions related to data privacy, ethics, and algorithmic biases. With AI being capable of storing vast amounts of personal information, what happens when someone hacks into the system? How can we manage data breaches to ensure that our information is safe? Some may even question the ethics of using AI and providing it with personal information in the first place. What exactly will this technology be used for? Malicious intent would certainly be a problem, especially if it is unclear who has access to this information. Even if these groups are identified, possibly as institutions such as governments, members of the public may still be skeptical of their intentions. So, what are the boundaries? Where do we draw the line of how we utilize AI? If AI usage does become a common occurrence in the retail industry, another issue could be potential algorithmic bias. For example, Amazon abandoned a tool that used AI to assess job applicants because the data used to develop the algorithm was based on previous applicants, who were predominantly men. As a result, the tool discriminated against female applicants (Davenport, Guha, Grewal, & Bressgott, 2019). In an age where discrimination within our society is actively sought to be destroyed, the biases in our technology could hold us back; ironic considering that robots may be seen as a progressive window into the future.

All of this is not to say that the widespread use of robotics in the retail industry would be good or bad. It is not exactly black-and-white; the concept of its application is not that straightforward. Robotics usage allows for smarter networks; since they are designed as part of an intelligent system, they can interact with other technologies as a cohesive unit to ensure that stores can operate efficiently through rapid communication and transfer of data. They positively impact economies of scale; a fleet of robots can support various business functions, thus reducing the average cost of ownership. The economic benefits of this technology are a massive pull for companies, and its gravity should only increase as those in the science field find ways to reduce costs. While human labourers may take a while to acquire and familiarize themselves with new skills through training, robots have comparatively easy upgradation of skills. Something as simple as downloading software can add a new set of skills to a robot's repertoire, thus improving performance capabilities in a short period of time. This enables them to perform their new job quicker than humans. The technological advances made by the scientific community have served the retail industry well, and wonders such as robots and artificial intelligence should be expected to become an even more integral part of company structures. In fact, a survey by Salesforce shows that AI will be the technology most adopted by marketers in the coming years (Davenport et al., 2019). On the contrary, automated agents increasingly will replace human employees; even in complex, analytical, intuitive, and empathetic tasks (Belanche et al., 2019). This will surely alter the landscape of the labour force and render many unemployed. Customers may be apprehensive to interact with robots, as this is a new concept. They may also perceive robots as less empathetic (Davenport et al., 2019). This would be particularly important in retail experiences that may require workers to be more emotionally in-tune with store patrons. The application of robots may invoke an issue of losing human connectedness. In addition, it was previously mentioned that robots could create a personalized retail experience based on a customer's preferences; but what happens if customers' preferences differ from their past selves? For instance, if someone wants to go on a diet, it then raises the issue of a robot presenting them with food choices reflecting on their previous lifestyle (Davenport et al., 2019). Robotics has so many benefits that we can and should take advantage of. They truly have the potential to enhance our lives. However, we should also examine and be wary of the potential risks and pitfalls that come with it. It may be best to monitor robotics applications so that balance can be found between these contrasting consequences. If we fail to do so in our attempts to propel humanity forward, we could very well push ourselves backwards.

Chapter 7

ROBOTICS AUTOMATION IN FOOD INDUSTRY

The advances in technology throughout the course of human history have given rise to robots: programmable devices consisting of electronic, electrical, or mechanical units that can manipulate or transport parts, tools, or specified manufacturing implements for industrial applications (Nayik, Muzaffar, & Gull, 2015). The capabilities of these machines allow businesses to automate processes, using these artificial beings to function in place of humans. Subsequently, robots have become the subject of desire for industries for various reasons; they do not fatigue like human workers; they can work in extreme physical conditions; they are more consistent when meeting specifications; they are more productive and efficient; they do not get bored with repetitive actions; they cannot be distracted; and in the food industry, they eliminate a possible contamination source (Nayik et al., 2015). Since manufacturers are constantly seeking ways to increase efficiency and output, the implementation of robotics automation has become a common answer. With so many benefits, it is a logical decision. But despite the seemingly endless list of advantages, there are pitfalls. While they are currently able to do wonders, most robots are not yet at a level at which they can execute certain complex tasks. Automation also raises questions surrounding the replacement of legions of workers and the removal of a large sector of jobs.

Regardless, the world is moving into an era of technological wonder - it is only fitting that robots become integral to the very industry that feeds us.

The early 1990s saw the first applications of robotics in the direct handling of food in the bakery industry. These machines were able to perform simple motor tasks, moving items from one place to another. They were able to do this at a reasonable rate of 55-80 cycles per minute (Nayik et al., 2015). Over time, many companies have grown to apply robots to various processes in many divisions of the food industry. Robotic packaging systems are utilized to place products into cartons or containers, unload food from cooking machinery, palletize beverages, and cap and label items.

Around the same time that baking started to involve robots in food handling, the dairy industry saw the introduction of Automatic Milking Systems (AMS). This technology was created in Europe and was implemented there in 1992, later becoming available in the US at the turn of the century. Each cow is fitted with an electronic tag that the robot uses to identify her. The cow enters the robot, which reads her tag and feeds her a reward based on her level of production. The robot cleans the cow's teats, attaches the milk cups, and begins the milking process. The cups disconnect when the milking is complete and the cow then exits the robot. Swedish company DeLaval developed the first commercially available robotic milking rotary at a farm in Australia. The rotary can milk up to 90 cows per hour and is composed of five robots; the first two clean and prepare the teats for milking, the third and fourth attach the milk cups to the teats, and the last one disinfects the teats before the cow exits. The robots use laser technology to determine the position of the teats. This method of milking has multiple benefits. Economically, there is no need to hire labour. It also removes the potential of human error and their lack of efficiency compared to robots. There is also an increase in milking frequency, leading to a higher milk yield per cow. The use of a robot computer can also lead to a/the more efficient management of the cattle herd (Nayik et al., 2015).

In the past twenty years, advances in various forms of technology have been immense. These developments have found their way into influencing agriculture (the practice of cultivating plants and livestock), where different technologies such as laser-based sensors and global positioning systems (GPS) have been incorporated into robots designed for the automation of agricultural tasks (Emmi, Gonzalez-De-Soto, Pajares, & Gonzalez-De-Santos, 2014). These systems provide accurate positioning and guidance for robots in the field, and thus with the right tools or equipment (fertilizers, seed planters, etc.), enable them to carry out precise tasks. For instance, the Gifu Information Technology Institute in Japan developed a weeding robot that operates autonomously in paddy fields (Kusuda, 2010). Such use of robots can greatly reduce or even eliminate the amount of time that workers and farmers spend on these jobs. The precision afforded by this technology also helps make them more productive and efficient than their human counterparts. Research groups are also trying to develop a system that can operate a fleet of vehicles under unified control (Emmi et al., 2014). The implementation of such a system in which the functions of robots are coordinated to carry out tasks will likely be an enticing notion for farmers with high-value and high-yield crops. However, the use of advanced technology still has its downsides. The addition of many systems increases costs and the greater complexity introduces more opportunities for things to go wrong. Nevertheless, as scientists find ways to reduce costs and make the technology more reliable, these concerns will be alleviated.

The food industry is one that has the potential to expose workers to extreme or even dangerous conditions. Produce, meat, dairy and other food items can become contaminated by pathogens such as e. Coli. The use of harsh chemicals such as pesticides brings more potential contaminants into the work environment that could be harmful to employees. So how can companies remove workers from harmful situations while maintaining similar or greater productivity? This is where robotics yet again comes into play. In the meat industry alone, there is a range of hazardous processes that robots can be used instead of humans. Butchering carcasses can be dangerous with the use of razor-sharp knives, as a lapse in concentration could cause serious injury for a worker. However, robots are able to operate these tools with complete precision, which can help reduce waste. Since robots cannot sustain injuries in the way humans do, there is no need to worry about a lack of production due to being in the hospital for multiple stitches. Working in meat freezers or storage boxes with low temperatures is not exactly beneficial for biological thermodynamics - robots can continue to work in these environments without having to step outside and take a break. While human employees can contaminate foods and beverages with viruses and bacteria, robots are much more sanitized and can be washed with more effective cleaning methods.

The costs of robots and the complex demands of dealing with biological materials has made the meat industry hesitant to jump onto the technological bandwagon, despite the obvious benefits of robots from a safety and hygiene perspective when it comes to tasks like the evisceration of animal carcasses. The Danish Meat Research Institute (DMRI) has had success in this area, collaborating with Danish company SFK-Danfotech to develop robots for automatic evisceration. This equipment can go through the processing and cleaning of 360 carcasses each hour; removing the pluck and intestines (Nayik et al., 2015). As technology becomes cheaper and more reliable to perform intricate tasks, more meat companies will likely adopt this trend and automation in this industry will start to become more common.

An estimated 76 million cases of foodborne illness occur annually in the United States of America. Most of these cases are the consequence of food contamination by handlers in processing facilities. Sick employees also can pose an contagious threat if they choose to either come into work or stay on the job while being ill. According to the Centers for Disease Control and Prevention (CDC), approximately 70% of all foodborne diseases in the United States are due to viruses that are spread through direct or indirect contact with those who are infected (Nayik et al., 2015). Either poor hygiene or illness can lead to employees spreading pathogens simply by handling food. Therefore, in a situation like the COVID-19 pandemic, sick workers can accelerate the spread of disease. One of the benefits of robotics automation is that it limits the contact between humans and food/beverage products to avoid the transmission of such germs. Additionally, to ensure that these devices (such as grippers for food handling robots) are thoroughly sanitized, they are washed with industrial agents and pressurized hot water (Iqbal, Khan, & Khalid, 2017). These thorough sanitation methods are often too harsh to be applied to humans, but they can safely be used on mechanical appendages to ensure that the point of contact for the food items is clean.

Robots managed to find their way into the front lines of the food industry. In the early 1990s, a restaurant in Okayama, Japan implemented robotic waiters. Since then, many other restaurants have followed their lead. A more recent example is the Hajime Robot Restaurant in Bangkok, Thailand, which utilizes robotic waiters dressed as samurai to serve customers. These machines also take customer interaction to a new level, as they also dance for entertainment (Kusuda, 2010). Going a step beyond serving food, robotics company AISEI opened a ramen restaurant, FA-Men, over a decade ago to promote their technology. Two robot chefs were employed to cook ramen noodle soup according to customer preferences before they were converted for industrial purposes in January of 2010 (Marx, 2010). Such technology may soon become common in our households in the future. Moley Robotics ([Moley], 2015) supposedly created a fully-automated and intelligent cooking robot that integrates the cooking skills of BBC Master Chef winner Tim Anderson. Moley (2015) claims that the system recorded Anderson's "every motion, nuance, and flourish." The consumer version is set to launch in 2020. The concept of a robot chef is truly phenomenal and seems like something out of a science fiction film, yet it has the potential to become normalized in the near future. The increased level of hygiene and precision that is incurred when robots are implemented are massive benefits, although some might argue that the human touch and spirit are integral parts of a food experience.

The introduction of robotics automation into the food industry yields substantial results and provides us with an abundance of benefits. It has made manufacturing processes more productive and decreased worker exposure to potentially harmful situations. It has also created a space-age feel for some restaurants. As these artificial systems become more intelligent and cost-effective, certain applications will become both possible and economically attainable. But how far should we go with it? As technology progresses to the point where robots can execute complex tasks as well as humans, they could easily render the living workforce as an unnecessary entity. Where do we draw the line between increasing automation and keeping workers on the payroll? In March 2020, just before the COVID-19 pandemic caused mass layoffs, there were over 1.2 million employees in accommodation and food services and around 300,000 employees in agriculture in Canada (Statistics Canada, 2020). At the rate that robots are becoming more complex, many of these workers may find themselves being replaced at some point in the future. This raises a different question; one that revolves around providing support for these workers. If most or all of the jobs in the food industry are automated, how will those who are unemployed find opportunities? The release of so many people into the job market could oversaturate it, and with the elimination of positions in the food industry, things may look bleak for workers whose skills can only be applied in this sector. It would be wise to try and avoid such a situation, and some solutions may help to do so. Workers could receive training to help develop their transferable skills, allowing them to better prepare for and fulfill positions in areas that cannot be dominated by machines. To avoid abrupt overflowing of the job market, the implementation of automation in the industry should be done in carefully planned long-term phases rather than in a short period of time. Governments may also want to provide some sort of financial support to those who become unemployed due to the use of robots. The potentially large influx of job-seekers could make obtaining employment difficult due to more competition, thus increasing the amount of time between jobs longer than the previous norm. Something similar to what governments have done with financial support in the face of COVID-19 could help families see through some tough times. Robotics automation has its upsides, and thus, a place in the food industry. However, we need to be wary of the cascading effects of implementing such technology.

Chapter 8

ARTIFICIAL INTELLIGENCE IN COVID-19 SCREENING AND PREVENTION

Due to the rampant spread of COVID-19, the urgency for artificial intelligence screening procedures is significantly increasing. Cutting-edge artificial intelligence screening procedures demonstrate the potential to introduce new methods to prevent the spread of the virus. It does this by detecting people who present the physical symptoms associated with COVID-19. Additionally, advanced AI screening will help prevent spread of the coronavirus. To explain, the current testing methods have significant wait times for results. Determinants of the delay in test results are primarily dependent on a person's place of residence, access to testing, the occurrence of outbreaks, and the number of tests administered in those particular areas. This means that individuals may be waiting on test results from anywhere between one day to over two weeks. The current screening measures are simply inadequate; therefore, the world must adapt and embrace new technologies in order to combat the spread of COVID-19, starting with increasing and innovating new forms of AI screening measures that can be utilized throughout the globe.

In order to contain COVID-19, screening for the virus is imperative. The first reported area affected by coronavirus, China, initially introduced the use of infrared imaging scanners as well as handheld thermometers in public locations. Over time, local artificial intelligence companies introduced more advanced, progressed, and effective AI technologies for screening. This includes AI temperature screening systems that are used in public spaces, like subway stations, in order to conduct mass screening. These technologies are capable of screening large public areas to test for signs of fever. Likewise, many regions across the world have introduced apps on smartphones that are AI-powered to monitor users' health and to track the geographical spread of coronavirus. These apps work to predict which areas are most prone to the negative impacts of viral outbreaks, allow people to receive current information from their healthcare providers, relay medical advice and updates, and notify users of infection hotspots. This AI technology is crucial as it gives people real-time information that they can use to distance themselves from virus hotspots and keeps those who are potentially ill away from public areas to protect themselves and others (Petropoulos, 2020).

Several Canadian companies have already developed new AI screening methods and procedures that will potentially overtake all other screening methods. For example, Predictmedix, an artificial intelligence firm from Toronto, Ontario, established screening advancements that will help numerous organizations to identify possible COVID-19 cases and other illnesses. Companies can utilize this adaptable innovation for the mass screening, scanning, and filtering of public spaces to monitor physical distancing and the use of face masks. This technology will detect any breaching of physical distancing measures and set off alarms to alert business employees to intervene. AI inventions demonstrate the potential for incorporation into a camera framework, of public and private areas, to screen for unsafe social behaviour (Patterson, 2020). Other Canadian companies, such as BlueDot, have also introduced new artificial intelligence screening technologies. BlueDot is an international leader for identifying emerging risks of COVID-19, and for predicting the international distribution of disease through the use of their own models. BlueDot conducts data-mining operations with large datasets to establish and provide current information regarding health services – this will show where the increasing demand is, as well as where critical equipment is needed (Sookman, 2020). Additionally, a Montreal-based artificial intelligence company, Nuvoola INC, also introduced their own unique form of COVID-19 screening, intended for businesses to limit viral transmission throughout their workplace. The system is called LUKE AI for Health Screening and Protection (HSP), and is composed of a touchless kiosk and an app. This app allows employees to submit daily reports on their health status before going to work. Once they arrive at work, employees must go to a kiosk where they are recognized by their employee identification or by facial recognition. The kiosk then screens people for COVID-19 symptoms and analyzes whether a mask is required, and then measures body temperature by using an infrared camera (Nuvoola, 2020).

One of the most effective prevention methods includes contact tracing. Contact tracing is vital in order to identify, monitor, and educate people who were in close contact with someone with the virus. This is especially important because people need to understand and learn to limit the risk of contracting and spreading the coronavirus. This AI technology and prevention method has already been enforced in many countries across the globe. For instance, an app called TraceTogether was launched in Singapore. It uses Bluetooth Relative Signal Strength Indicators (RRSI) reading between devices in order to estimate and record encounters between users. This information is stored on a person's phone for three weeks, but no location data is collected. However, when someone downloads this app, they are required by law to assist the Ministry of Health (MOH) to create a record of their actions over the last 14 days. This includes the possible collection of documents, records, or data from other apps on the user's device. Then, if a user of the app becomes ill with COVID-19, contact tracing would activate. This means that the MOH would then contact the individual to identify and retrieve information that outlines the individual's activity over the last 14 days (Sookman, 2020).

As mentioned in a research article written by Ferretti and colleagues (2020) regarding the quantification of COVID-19 transmission through AI contact tracing, until there is an uncomplicated access and availability of a vaccine, there are only a few approaches to infection prevention. These approaches include physical distancing, decontamination, hygiene measures, case isolation, contact tracing, and quarantine. Likewise, there is a contact-tracing smartphone app issued by the Government of Canada that stores information of people who report a diagnosis of coronavirus. This app runs in the background of its user's phone, and uses bluetooth networks to exchange information with other smartphone users. If a person comes into close contact with someone infected with COVID-19, then the app will notify the person and suggest that they get tested for the virus (Canada, 2020). This form of AI is extremely beneficial as it allows for everyone who has been exposed to the virus to be notified on a broader scale, and prevent future transmissions of COVID-19. Therefore, this strongly indicates that it is vital to implement contact tracing as a form of prevention in order to combat the spread of COVID-19 and protect people. Not only does this help to protect average citizens, but it will protect all populations—vulnerable people, residents living in remote areas, and minority populations such as Indigenous peoples.

Even though artificial intelligence in COVID-19 screening appears to be the most called-for innovation during the current predicament, it does bring forth some public concerns. One of the biggest downfalls of using AI is the liability of user privacy. This is due to the fact that AI software collects and stores information, which raises issues under the Canadian Charter of Rights and Freedoms (CCRF). These public surveillance technologies use location information stored on smartphones, facial recognition, public-area infrared scanning for fever detection, and other computer surveillance technologies (Sookman, 2020). They are also dependent on access to smartphones or other technologies that may not be available to the majority of the population. On a similar note, these technologies can be quite expensive to produce and may not be ideal or affordable to implement with the accumulating debt worldwide. However, risks need to be taken to some degree in order to limit the spread of COVID-19.

The impact of the novel coronavirus tremendously impacts the people living in remote areas, vulnerable populations, and Indigenous peoples in Canada. These groups are particularly susceptible to the virus and suffer the negative effects of trying to recover some semblance of normalcy in a world overrun by coronavirus. As the threat of the coronavirus continues, so does the impact it has on these communities. People living in remote areas are vulnerable as not many resources are available for when someone contracts the virus. They have limited access to both human and medical resources; nurses and doctors fly in very few times annually, making it difficult for these communities to combat an outbreak of this scale. Due to the large population sizes in these areas, residents have inadequate equipment or treatment methods in their communities, forcing ill individuals to travel elsewhere to receive treatment. This can be very risky as travel promotes viral transmission of COVID-19. The lack of medical resources is very unsafe for these populations as it means that if someone develops the virus and it spreads throughout the population, there likely would not be enough resources to support the ill populations. Likewise, any deliveries made to these communities create a highly volatile situation for coronavirus transmission. Members of immunocompromised, elderly, and low-income populations are at high-risk for COVID-19 and they often cannot afford to shop online, buy adequate food, or purchase medications. This already makes it very challenging for them to provide for themselves, so if they were to contract the virus, it could potentially lead to serious or fatal outcomes. Additionally, Indigenous peoples are also at increased risks for COVID-19 due to several health reasons. First, the prevalence of diabetes among Indigenous populations is at alarming levels. According to the United Nations, type 3 diabetes occurs in more than 50 percent of Indigenous adults, and these numbers will likely rise (United Nations [UN], n.d.). Indigenous communities often suffer from a lower quality of life, which results in much poorer health outcomes compared to their non-Indigenous counterparts. Altogether, Indigenous peoples experience insufficient medical care more frequently than any other population; an issue which has especially taken its toll on Indigenous women. From a health perspective, Indigenous women demonstrate increased cases of maternal and infant mortality, infectious diseases, malnutrition, cardiovascular illnesses, and many other diseases and illnesses (UN, n.d.). Due to the prevalence of these health issues, it makes them very susceptible to complications of the virus. The reserves Indigenous peoples live on often have inadequate accommodation and maintenance, unclean water, inadequate medication, less travel access, and poor medical resources. All of these populations are very high-risk; therefore, people must do as much as they can to protect these communities in order to limit the number of fatalities as a result of the virus.

People across the world must continue to develop artificial intelligence technologies in order to limit the spread of COVID-19. The current testing measures, such as the use of RT-PCR, are simply inadequate to accommodate the severity and rapid dispersion of this virus. More immediate methods of testing must be created so that people can have their results instantaneously, thus making it easier for someone to self-isolate and limit the spread as fast as possible. The world needs more instant testing in order to combat this virus. That is why artificial intelligence screening methods must become more prevalent so that it can keep as many people safe as possible. Canadian companies previously mentioned, such as Predictmedix, BlueDot, and Nuvoola, are doing a great job at speeding up the process to fight this virus and provide as much safety as possible to people worldwide. Even though artificial intelligence will not be able to defeat coronavirus on its own, it provides many benefits that would significantly speed up the process. Humans can only perform tasks at a certain speed, but artificial intelligence allows for this speed to increase and produce more efficient results. It leaves less room for human error, and consequently, it can improve results from screening methods and also increase prevention measures. AI companies must continue to fabricate technological advances in order to provide maximal protection to people. The world must continue to make advancements regarding coronavirus prevention and screening in order to gain control of the pandemic, and ultimately put this worldwide pandemic to a halt.

Chapter 9

ARTIFICIAL INTELLIGENCE IN COVID-19 RESEARCH AND TREATMENT

The use of artificial intelligence (AI) is vital in the research and development of effective treatment methods for COVID-19 patients worldwide. Artificial intelligence offers highly advanced capabilities to aid researchers to filter and analyze massive volumes of research data in a fraction of time typically needed. AI technology helps tailor the research in a direction that is the most effective for meeting the researcher's goals. The ability for artificial intelligence to efficiently streamline significant amounts of data is particularly useful in this time-sensitive situation, with scientists worldwide scrambling to put an end to this global epidemic. Therefore, researchers are utilizing AI-powered technology and machine learning in order to discover potential safe and effective treatments for COVID-19, as well as methods to help countries across the world recover from this life-changing pandemic. AI technology assists researchers to view things from a health-care perspective, and provide relevant information that can help the world work towards reconciling their social, economic, and psychological health needs (Kent, 2020).

First, artificial intelligence (AI) technologies enable researchers to take impactful steps toward COVID-19 research. Artificial intelligence can filter through wide-ranging data to find, evaluate, and summarize thousands of coronavirus research papers. Interestingly, AI detects patterns in vast sums of data that people often fail to recognize—it provides a whole new perspective that helps researchers to gain a new understanding of the data (Fung, 2020). The use of artificial intelligence is crucial in gathering data in a time-sensitive manner. The typical turnaround time for human predictive models is 314 days, but with the new AI technology, scientists can effectively shorten the turnaround time to accomplish the same research in only five minutes (Stoner, 2020).

Since institutions across the globe are concurrently researching this virus, scrambling to find treatments, artificial intelligence can produce a beneficial summary of all of the findings produced by countless researchers. As more research surrounding the coronavirus comes to light, scientists can determine which drugs and treatments are best suited for different populations and sub-populations of patients. Artificial intelligence systems could also speed up this process thanks to their ability to detect different customized therapeutic targets for various communities. In turn, researchers could use these information data systems to create treatment algorithms and guidelines customized for patients with any symptom from the broad spectrum of coronavirus symptoms (Stoner, 2020).

As scientists across the globe hope to find some form of cure or treatment for COVID-19, many are seeking assistance from another type of artificial intelligence called machine learning. Machine learning enables computers to analyze large amounts of data, identify patterns and insights, and essentially mimic a more advanced version of human intelligence. In addition, artificial intelligence identifies vulnerable populations and acts as an early warning system for future populations. For instance, researchers at the Chan Zuckerberg Biohub in California built a model that works to approximate the number of undetected COVID-19 cases, and evaluate the consequences on public health. They are doing so by analyzing different characteristics of the virus, including how it mutates as it spreads through populations, and estimating the number of missed reports of transmission (Sivasubramanian, 2020).

BenevolentAI is a company from the U.K. that hopes to integrate machine learning in the fight against COVID-19, by remodeling it as a discovery platform for drug treatments with the potential to inhibit the progression of the coronavirus. More specifically, BenevolentAI used this technology to identify relationships between genes, diseases and drugs, to narrow down many medications into a small number of drug compounds. It is important to reduce the number of drug compounds as it brings scientists closer to developing an effective treatment for the virus. In fact, through machine learning, BenevolentAI discovered that Baricitinib, a drug used for rheumatoid arthritis, is a promising candidate for a drug treatment for COVID-19. As Baricitinib has now entered its final stages of clinical trial, the effectiveness of machine learning in accelerating the research project becomes apparent. Therefore, the use of machine learning to detect drug compounds is a crucial resource for the current pandemic and it could potentially discover the cure for COVID-19 (Sivasubramanian, 2020).

Currently, there are many active, COVID-related projects that have begun exploring the use of artificial intelligence. For example, a team at Penn State's Institute for Computational and Data Sciences (ICDS) is currently utilizing an artificial intelligence supercomputer to find potential treatments for COVID-19 (Kent, 2020). Likewise, the ICDS supercomputer, named ICDS-ACI, is also helping members from the Institute's Research Innovations with Scientists and Engineers (RISE) to analyze large quantities of data. As COVID-19 is the current hot-topic in research, ICDS-ACI's data-handling capabilities is imperative to simplify the tremendous amounts of data. More specifically, a task that may take the typical laptop days to complete—this supercomputer can complete it within seconds to minutes. ICDS has granted access to its supercomputer for outside researchers to aid them in their projects and research. Examples of these projects include the study of wireless communication modules in the form of a wearable chest patch. This allows medical professionals to closely observe isolated coronavirus patients without breaching social distancing measures. As beneficial as supercomputers are in the lab, they come at an expensive price, making it important for researchers to seek financial support along the process. Therefore, these grants play a decisive role in the process of treatment research for COVID-19 (Kent, 2020).

Another project involving artificial intelligence includes researchers Jean-Philippe Julien and Costin Antonescu (Fraumeni, 2020). Jean-Philipe Julien is an associate professor in the department of biochemistry and immunology at the University of Toronto, as well as a senior scientist at the Hospital for Sick Children. Constin Antonescu is an associate professor at Ryerson University, who also completed his Ph.D. at the University of Toronto. These two researchers are shifting their focus towards existing drugs, including antiviral drugs, to see if they have the potential to treat coronavirus. In doing so, they are using artificial intelligence which is modelled based on neurons in the human brain. This form of AI relies on neural networks to learn, comprehend information, and make decisions. Although their study is currently ongoing, the pair has already found molecules that show the potential to prevent the virus from entering cells in a human body, which perfectly showcases the promising capabilities of AI in scientific research.

Another group of people that uses artificial intelligence to fuel their COVID-19 research is Northwestern University in Evanston, Illinois. Similar to previous examples, this research-intensive university is applying artificial intelligence (AI) to improve the search for treatments for the virus. This technology, CAVIDOTS (Coronavirus Document Text Summarization), allows them to make faster decisions amid the search for cures. The user can operate the software by visiting a website application and inputting relevant search terms. They can choose to begin their search with broader categories and eventually narrow it down to more specific keywords. CAVIDOTS takes these keywords and scans through scholarly articles found on the COVID-19 Research Dataset (CORD-19). CAVIDOTS then predicts the most useful results for the user, and it generates a summary for the researchers (Morris, 2020). According to Brian Uzzi, a professor at Northwestern University, "The standard process is too expensive, both financially and in terms of opportunity costs. First, it takes too long to move on to the second phase of testing, and second, when experts are spending their time reviewing other people's work, it means they are not in the lab conducting their own research." Uzzi, along with other researchers from Northwestern University, developed an algorithm to be used as a method of predicting the study results that are most likely to be reproduced with new populations (Stoner, 2020). The algorithm indicates which studies' results are most promising and most likely to prove effectiveness. According to researchers, the replicability prediction models of these machines are potentially more accurate than the regular prediction markets; AI takes the narrative of the study into account, whereas human expert reviewers often direct their attention to the relational statistics in a research paper. Since the machine's algorithm inspects and reviews words in thousands of papers, it begins to recognize distinct word-choice patterns that humans are often oblivious to. This technology builds upon and expands the capacity of human intelligence, which makes it an outstanding tool to couple with human reviewers (Stoner, 2020).

One shortfall to AI machine learning and treatment is its inability to cater to all elements of the human condition. Not only does the human condition describe the entirety of living, but it includes everything that composes a human ("Human-condition," n.d.), which is why it's so important that AI accommodates all of these factors. Unless the AI system is programmed to account for particular criteria or factors, the results may exclude entire sectors of the population (Davidson, 2020). The exclusion of sub-populations is an issue that has come up in indigenous communities across Canada. Canada is well-known for having a diverse and dispersed indigenous population, and these communities continuously face systemic discrimination by other Canadian systems and policies. A widespread concern—particularly prominent among indigenous communities—is that the AI programs are created by people who may not be aware of the specific challenges facing that community (Davidson, 2020). For example, many indigenous peoples do not have access to clean water, proper health care services, and even basic human rights. Likewise, indigenous populations in Canada generally have higher rates of underlying health conditions such as tuberculosis and diabetes. These diseases put them at higher risk for COVID-19, and are primarily associated with poverty and inequity. According to the World Health Organization, people diagnosed with both coronavirus and tuberculosis often have worse outcomes with treatment, especially if their treatment for tuberculosis is interrupted ("PAHO/WHO," n.d.). Amanda Carling (Carling & Mankani, 2020), the manager of Indigenous Initiatives at the University of Toronto, added that the indigenous populations are often faced with discrimination when attempting to access any form of health care service. Given the fact that COVID-19 is spreading across the world and forcing travel restrictions, it is making travel to medical centres extremely difficult for indigenous peoples. Not only is it nearly inaccessible, but it is expensive. As previously mentioned, this issue is especially prominent since many indigenous communities and researches are in remote locations. Numerous indigenous communities are experiencing limited access to the proper forms of personal protective equipment (PPE) that they need. Carling also mentioned that these circumstances do not even slightly justify the harmful mental health related consequences that members of these communities will face in the post-pandemic future (Carling & Mankani, 2020). Living conditions are already quite challenging for the indigenous populations, but the novel coronavirus is making life a lot more challenging for these communities.

In conclusion, given the severity of the novel coronavirus, researchers across the globe are still far from defeating the virus. We must utilize different resources, such as artificial intelligence, to find and streamline as much information as possible regarding treatment options. It is also crucial that the AI programs are comprehensive in their data collection to not overlook the issues that can affect specific sub-populations. AI technology can be an instrumental tool used to help researchers determine effective treatment options in the most time-sensitive ways as possible. It is imperative for people with access to AI technology to utilize it in a way that maximizes the benefit for all people. With high hopes, AI technologies hold the potential to lead us out of this coronavirus shadow into recovery, assisting us in our journey as we build resilience and ready ourselves for future uncertainties.

Chapter 10

ROBOTS USED IN THE HOSPITAL DURING COVID-19

Robots are powerful tools that serve many purposes in the healthcare system. They assist with surgical procedures, deliver medical treatment, provide patient service, and maintain clinical environments. Robot-assisted surgery (RAS) is a practical alternative for medical professionals to use during these unprecedented times. RAS reduces the risk of transmission by: lessening the duration of a patient's hospital stay, limiting direct human contact, and minimizes exposure to bodily fluids (Mottrie, n.d.). These outcomes create more room to accomodate incoming patients, which consequently lessens the burden on healthcare workers. Likewise, other robots disinfect the building, replacing caretaking staff. This is beneficial for the staff as they can limit their chances of contracting the virus. Therefore, robots demonstrate noteworthy potential in hospitals during the COVID-19 pandemic.

Robot-assisted surgery (RAS) is beneficial for many reasons. The sudden influx of COVID-related hospital admittances forces hospitals to prioritize their patients—the first concern is COVID patients, while others are put on the back burner. This generally results in the suspension of discretionary surgeries to make room for COVID patients in operating rooms, intensive care units, and recovery rooms (Kimmig et al., 2020). While most hospitals have set non-emergent surgeries aside, certain surgeries cannot be significantly delayed, which places an immense burden on the healthcare system as it struggles to withstand the rapid influx of patients. This is where the benefits of RAS become apparent. To begin, RAS minimizes the number of operating staff required to perform the procedure. Surgeons can operate on patients remotely through teleoperation, which is when someone operates a machine from a distance. The surgeon uses master controls to navigate the robot to make precise incisions and movements inside the patient's body (Health, n.d.). Aside from assisting inside the operating room, RAS also proves to be beneficial for postoperative recovery and minimizing hospital stay. Surgeries performed by robots often demonstrate more precision, therefore leaving patients with smaller surgical wounds that take less time to heal. Faster healing indicates faster discharge from the hospital post-operation. By reducing the patient's duration of stay in a high-risk environment, it effectively minimizes their exposure to the virus and limits viral transmission. Another benefit that comes with shortened hospital stay is increased hospital turnover. This prepares hospitals for more overcrowding in the future as COVID-19 cases continue to surge. Consequently, the reduction in time increases the hospital's turnover, thus minimizing the transmission risk between patients and staff. In standard surgery methods such as open or conventional laparoscopic surgery, there is a greater risk of contamination with body fluids and surgical gasses (Kimmig et al., 2020), which is why it is important to use RAS to minimize as many risks as possible.

In order for the surgeries to occur, the surgeons are given important recommendations, especially on patients who are COVID-19 positive. This includes general protection and precautions, surgical technique, prevention and management of aerosol dispersal, pneumoperitoneum disinflation, and transmission through the patient's bodily fluids. In the stage of general protection, patients undergo preoperative screening, regardless of being symptomatic or asymptomatic (Mottrie, n.d.). Preoperative screening must take place in order to analyze the urgency and necessity of surgery. If a patient tested positive for COVID-19, and does not exhibit life-threatening conditions, then doctors are expected to postpone the surgery. However, in the case of emergency surgery, doctors must perform the procedures in a designated operating room (OR) and follow clinical protocols regarding the protection of operating room staff. Even if the patient tested negative for COVID-19, practitioners must consider the possibilities of false negatives. This means that all of the necessary precautions must still take place to minimize the risk of transmission (Mottrie, n.d.). These precautions are as follows: first, before any surgical procedures can take place, the surgical team must fulfill certain prerequisites. Only experienced surgeons should perform surgeries to prevent human error and eliminate potential COVID-19 transmission risks (Ficarra et al., 2020). Next, with regards to personal protection requirements, it is pivotal that bedside assistants fulfill the level III protection standards. This level consists of wearing personal protective equipment (PPE) including a disposable surgical cap, disposable latex gloves, disposable medical protective uniform, medical protective mask (FFP3) and goggles or visor. A full-coverage respiratory protective device or powered air-purifying respirator is also preferred. The console surgeon—or the surgeon responsible for operating the controls from a distance—must take level II precautions. Taking a step down from level III, this involves the same set of PPE of a disposable surgical cap, disposable latex gloves, disposable medical protective uniform, medical protective mask (FFP3) and goggles or visor (Kimmig, et al. 2020). The only difference is that level II is only air purifying with dermal protection, whereas level III is the full spectrum of PPE ("Personal," n.d.).

Prevention and management of aerosol dispersal are also crucial areas of consideration during surgical procedures. All surgical instruments must be kept clean of blood and bodily fluids (Zheng et al., 2020). During most surgeries, medical professionals use devices such as lasers, electrosurgical units, ultrasonic devices, and high-speed burrs, drills and saws. Although these are all effective surgical tools, they often result in the creation of gaseous by-products such as surgical smoke (Y. Liu et al., 2019). Research has shown that surgical smoke can put surgeons, nurses, technicians, and anesthesiologists at risk for medical issues. This is because the release of surgical smoke contains chemicals, bacteria, viruses, blood particles, and tissue particles (Y. Liu et al., 2019), which can potentially carry small viral particles (Mottrie, n.d.). In order to manage and prevent aerosol dispersal, medical staff should specifically monitor things such as pneumoperitoneum, hemostasis, and the cleaning of incisions to prevent the leakage of bodily fluids (Zheng et al., 2020). They must ensure that they fulfill the optimal effectiveness and safety measures in the gas evacuation. It is also important to reduce the intra-abdominal pressure to less than or equal to 8mmHg (Mottrie, n.d.)., and they must also ensure the disinflation of the pneumoperitoneum was conducted correctly

(Y. Liu et al., 2019). Pneumoperitoneum describes the presence of gas or air in a person's peritoneal cavity, which is found in the abdomen (Sureka et al., 2015). If the medical staff fails to do so, they are at risk of exposure to the patient's pneumoperitoneum aerosol. This commonly occurs when there is a sudden release of the trocar valves, if they fail to conduct an airtight exchange of instruments, or through small incisions in the abdomen. Conversely, surgeons often use surgical insufflation during procedures. This safely transports carbon dioxide into the abdomen to essentially inflate the surgical area and enable improved access to anatomy ("Insufflation," 2018).

Researchers discovered that SARS-CoV-2 is present in human stool. Therefore, this identifies a potential risk of transmission of SARS-CoV-2 through the handling of human feces. Medical staff need to ensure that they are avoiding contact with human bowels as much as possible to minimize the risk of transmission of COVID-19 (Mottrie, n.d.). Similarly, researchers also concluded that since there has been a continuous detection of viral nucleic acids in feces, this suggests that COVID-19 may be transmitted throughout the digestive tract or even re-transmitted through coronavirus-containing aerosols (Ling et al., 2020).

A study also discovered that there is a possible presence of SARS-CoV-2 nucleic acid in human urine (Ling et al., 2020). Although the data did not illustrate a direct correlation in robotic procedures between urine spillage and virus transmission, surgeons should still proceed with caution as there is no concrete evidence at the moment (Mottrie, n.d.).

In the case of renal transplantations, practitioners must take several precautions. More specifically, transplantation should only occur if it is urgent and life-threatening for the patient. In order to avoid transmission of the coronavirus, both donors and recipients must be tested COVID-19 negative before the surgery occurs (Mottrie, n.d.). There is currently limited information available regarding transplant surgeries and COVID-19; therefore, strict precautions should still take place in order to avoid the transmission of SARS-CoV-2 as much as possible.

Throughout the hospital, robots are very useful for a variety of tasks. Even though the use of robots is not yet universal, it is increasing in popularity as more hospitals adapt to this new technology. UVD Robots, a company based in Odense, Denmark, created robots and sent hundreds of them to frontline workers across the world. More specifically, they sent out disinfection robots to places such as Wuhan, Rome, and Veneto. The robots can decontaminate surfaces by emitting powerful ultraviolet (UV) light that tears apart the strands of viral DNA. They will first analyze and map out their environment, then move around autonomously while shining its 360-degree UV-C light. This is ideal for disinfecting vast areas in public places, particularly hospitals as it protects staff and patients by minimizing human exposure (Hornyak, 2020). Xenex Disinfection Services, a company based out of Texas, also created a disinfection robot called the LightStrike Robots. LightStrike Robots use pulsed xenon to create UV light and therefore disinfect areas (Hornyak, 2020). This is ideal for hospital use as they reduce the time used to disinfect surfaces as well as the amount of personal protective equipment (PPE) required (Murphy, Adams, & Gandudi, 2020). They also demonstrate more efficiency when cleaning compared to human cleaners. Therefore, it appears to be the safest option given the prevalence of COVID-19 across the world.

During the coronavirus outbreak, the use of robots in hospitals must be accessible and adopted internationally. Not only do these robots lessen the burden on healthcare workers, but they provide more protection for the staff and patients. Developing these styles of robots will also provide an easy, well-planned solution in the eventuality of a second wave of the virus. For example, in March 2020, a hospital staffed entirely by robots opened in Wuhan, China. This hospital was named the Smart Field Hospital, which was previously the Hongshan Sports Center (Hornyak, 2020). This project involves Wuhan Wuchang Hospital, China Mobile, CloudMinds, and the Chinese Academy of Sciences Research Institute (Houser, 2020). This Smart Field Hospital is acting as a trial to find a method of relieving healthcare workers who are exhausted as a result of coronavirus. It allows 12 sets of robots as well as other IoT devices to treat patients who are showing mild symptoms of SARS-CoV-2 (Houser, 2020). When patients enter the hospital, they are screened for coronavirus through the use of 5G thermometers that alert staff members if there are any feverish people. Upon entering, the patients are given a smart bracelet and ring that would sync with the artificial intelligence (AI) platform. The staff members were also required to wear these to monitor any early signs of infection of the coronavirus. The CloudMinds' AI platform used these devices to monitor a patient's vital signs, including temperature, blood oxygen levels, and heart rate. The other robots had duties such as food and drink delivery, medicine delivery, decontamination and disinfection, disposal of medical waste, and entertainment (Hornyak, 2020). Not only did these robots limit the transmission risk of COVID-19, but it also provided a mental health benefit for its patients. In quarantine, many feel isolated and lonely; however, these robots provided them with entertainment and connections to boost their mood. Due to the fact that these robots are teleoperated, previous health care workers' jobs were not in jeopardy – they controlled the robots so that they could still apply their expertise and demonstrate compassion for the hospital patients (Hornyak, 2020).

Telemedicine, or telehealth, is the use of electronic communication so that people can receive remote health care. Telemedicine is not a newly-discovered technology – it has been around for years. However, given the emergence of the novel coronavirus, telemedicine is in high demand. It is quickly shifting the paradigm regarding how patients receive healthcare.

Hospitals across the world, such as Rush University Medical Center in Chicago and George Washington University Hospital in Washington, D.C., introduced the use of telemedicine to assist in the screening and detection of COVID-19 among their patients (Hornyak, 2020). Canada encourages its provinces to practice telemedicine and virtual care whenever possible during the pandemic ("Telemedicine and virtual care guidelines," n.d.).

Additionally, a Minneapolis-based company called Zipnosis created online questionnaires to be completed by patients. Doctors receive the results to review - they must decide whether the patient was exhibiting symptoms of coronavirus or symptoms from other respiratory illnesses. This is a very useful tool as it separates probable COVID-19 cases from other illnesses, thus limiting the risk of transmission. During the 2017 measles outbreak in the United States, people also used this platform as it took approximately 1 minute and 29 seconds to assess a patient's medical condition (Hornyak, 2020). This significantly cuts down any wait-times for medical attention and ensures the satisfaction of people's medical needs. Unfortunately, this service may be inaccurate - there is no way to properly ensure a patient's honesty of their symptoms. Likewise, proper medical testing cannot occur through telemedicine. In order to get a definite diagnosis, patients would have to undergo medical testing in an office or lab. However, this platform still provides a method to meet medical demands during the pandemic as there is a colossal lack of available resources for patients and medical professionals.

In conclusion, the use of robots in hospital settings during the COVID-19 pandemic shows promising results. Whether it be through robot-assisted surgery (RAS), treatment of patients with robots, or general upkeep throughout the hospital, they all offer advantages to the healthcare system. Robots help to limit human exposure, hospital-stay times, opens up room for COVID-19 patients, reduces contamination, and most importantly, reduces the risk for virus transmission. Robots are becoming an essential tool that can be programmed to optimize its usefulness in the hospital setting. It is evident that coronavirus is significantly impacting the world – many people are hospitalized or even lose their lives due to the virus. Therefore, humans must do everything in their power to discover the most efficient and effective methods and technologies in their fight against the novel coronavirus.

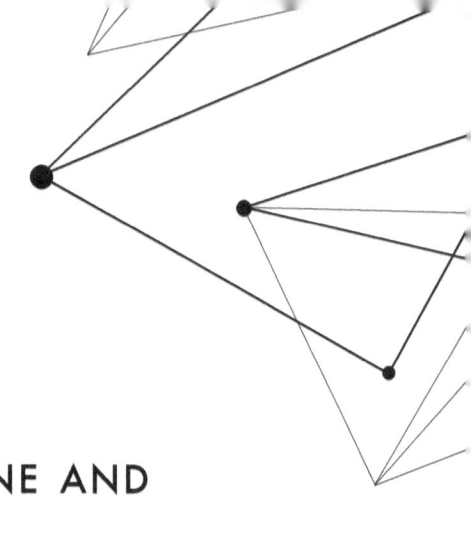

Chapter 11

ROBOTS IN QUARANTINE AND NURSING HOMES

The coronavirus pandemic is significantly impacting the lives of people across the world. Our typical lifestyles of going to work, grabbing a bite to eat with friends, and going shopping, shifted to days spent inside, cut off from the chaos outside our doors. We are quickly adapting to a life consisting of social distancing, self-isolation, wearing masks, and religiously using Lysol wipes. Scientists, researchers, entrepreneurs, and many other professionals struggle to find new and safe ways to adapt to the drastic changes happening across the world. In fact, robots are responsible for a significant shift in the way that the world carries out tasks. Self-isolated families and seniors in long-term care (LTC) homes incorporate the use of robots and drones into their daily lives. People are desperately reaching for new ways to creatively use robots to lessen the heavy burden on us.

Robots play an active role both in quarantine and in public settings. Not only do they monitor those living in quarantine, but robots assure public safety by enforcing proper social distancing measures. In communal areas, drones observe crowds, ensuring that no one is disobeying their quarantine restrictions. The thermal imagery system of drones identifies any cases of fever and relays this information to public health authorities. Using this information, local authorities can ensure the safety of other citizens by directing them away from coronavirus hotspots (Murphy et al., 2020). As the pandemic unfolds, workplaces are beginning to integrate robots into their company's routines. Realtors are socially distancing from their clients by using robots to conduct property tours while remaining in the comfort of their own homes. Companies in China utilize drones to provide lighting for construction workers to build hospitals through the darker hours. Schools in Japan use robots to mandate physical distancing measures so that graduating students can still walk the stage and receive their diplomas. Taking a closer look at the integration of robots in society, a company called CloudMinds has recently created Smart Transportation Robots, Cloud Ginger (aka XR-1), to deliver food, medicine, and other essential supplies to those in quarantine without direct human contact. XR-1 provides its users with continuous information, interaction, and entertainment (Hornyak, 2020). Bill Huang, the CEO and Founder of CloudMinds, mentioned that the patients found humanoid robots, such as Cloud Ginger, to be very helpful during these troubling times. Another group company based out of Japan, Terra Drone and Antwork, created a robot similar to Cloud Ginger. During the peak of the pandemic, Terra Drone and Antwork's robot transported medical samples and quarantine materials throughout Xinchang, China (Hornyak, 2020). Different areas in the world, primarily China, demonstrated during the height of the pandemic that robots are critical resources to make use of to minimize the amount of human contact.

As stated in the previous chapter, healthcare professionals operate robots to optimize the safety in healthcare. Doctors send the robot into the rooms of their patients, where the robot assesses the patient's breathing, blood pressure, vitals, and many other symptoms, without having to come in direct contact with the ill patient. Not only does this minimize the risk of contracting COVID-19, but it also decreases the demand for personal protective equipment (PPE), which is vital in situations where hospitals face shortages (Krauth, 2020). Many different companies are taking on projects involving the use of robots and supercomputers in telemedicine. For instance, the ICDS Explore Grant program funds several projects that explore the use of supercomputers. One of the grants funds researchers who are designing a wireless communication module in the form of a chest patch, which analyzes patients in quarantine and detects if any of them are in critical condition. Monitoring the quarantined patients with this chest patch enables patients to remain under vigilant observation while also preventing patients from transmitting the virus. Likewise, the wireless communication patch lessens the burden on hospitals currently operating at maximum capacity ("Robots Help Elderly Speak," 2020). Knowing that we must reduce physical contact and limit viral transmission as much as possible, these robotic solutions provide a much safer alternative for administering medical treatment to individuals in quarantine to maximize our safety.

Quarantine is putting a wrench into many people's lives, not excluding elders living in long term care (LTC) homes. With the implementation of countless new restrictions globally, nursing homes are also putting new COVID-19 precautions, like limiting the number of visitors at the homes. Due to the recurrent shortages of PPE, and the occasional workers, it is becoming quite challenging for LTC homes to provide adequate support for their elderly residents (Guerrini, 2020). It does not help that large volumes of outbreaks have occurred in LTC homes across Canada, leaving many residents and staff with serious infections. Unfortunately, as the elderly tend to have weaker immune systems, a large portion of these infected people pass away. Employees from the Belgian robotics software, also known as Zorabots, realized they had an excess of stock due to the worldwide outbreak of coronavirus. The company decided that one of the ultimate ways to put the abundance of materials to use is by giving it to the elderly communities ("Robots Help Elderly Speak," 2020). These communities needed help, so this was the perfect solution to utilize their abundance of resources. Caregiving is a very physically demanding job. Caregivers repeatedly lift elderly residents out of their beds, help them shower, and help with any other task that is hard for the seniors to perform on their own. Understandably, these actions can lead to the development of many physical issues such as back pain. A good portion of the robots distributed in Japan aid caregivers with these tiring tasks, alleviating their burdens while working towards a common goal: to provide the elderly with the best care possible during this period of adversity ("Robots May Be," 2020).

Not only do robots help the elderly with physical tasks, but they also assist with their mental and cognitive well-being. Over the last couple of years, scientists have been researching to create computer-based programs for the elderly. Scientists converted the programs into a robotic form, with the intention of using cognitive games to maintain and improve the senior's cognitive state (Coşar et al., 2020). In a project conducted by SARA Robotics, they created a robot – the SARA robot – as a part of the "Social & Autonomous Robotic Health Assistant" program, which gained attention just last year at a conference in Brussels, Belgium. The purpose of this robot is to improve the quality of life for senior citizens by providing opportunities to play interactive games, listen to music, and many other applications (Guerrini, 2020). Additionally, SARA helps first-stage dementia patients by offering cognitive games for mental stimulation. After SARA gained recognition, many users recommended that the SARA robot introduce ways to contact their families while they are stuck inside the nursing home. Given the request, SARA Robotics added the option to share videos and photos, and eventually, incorporated a way to video call people outside of the long-term care home (Guerrini, 2020). Another study investigated the effects of robots on the social and emotional needs of seniors. Long-term care homes received a seal robot named Paro to try to entice and engage the seniors. Through the study, they discovered that Paro was able to continuously uplift the feelings of the senior citizens who were actively engaging with it. The researchers also determined in another study that seniors who interacted with the robot in a group were more likely to socialize with others in the home. Paro essentially served as the catalyst for conversation at these long-term care homes (Guerrini, 2020). Even if these projects do not provide immediate change, hopefully, they can fill the current gaps in the current system of elderly care. These attributes help to improve the mental, emotional, and physical fitness of the senior, which is crucial during this unprecedented time. Quarantine measures often restrict in-person visits, but robots enable people to communicate and interact in ways that cheer them up. Robotics provides a useful means to help eliminate senior isolation; seniors need to receive lots of attention and continue to socialize with their families and peers. Due to human ignorance and oblivion, people often overlook elderly depression. The ignorance gives us even more of a reason to take extra care of them, especially during these difficult, uncertain times.

The coronavirus has taken an indisputable toll on the mental and social well-being of people across the globe. Many people will likely carry this psychological trauma with them for years to come, whether it be mild or severe. Common effects of living in the coronavirus pandemic include anxiety, uncertainty, fear, chronic stress, social isolation, and economic difficulties.

People feel exhausted in what feels like a never-ending catastrophe (Vitelli, 2020). People across the world, especially those in isolation, are realizing just how important it is to have social connections. Often, with social isolation, there is an increased risk for isolated individuals to develop suicidal thoughts and behaviours. Unfortunately, the majority of current preventative measures for coronavirus include social distance, which consequently increases the risk of developing these suicidal ideations (Reger et al., 2020). Due to public safety measures, social distancing, and social isolation, it sets up barriers for individuals seeking mental health treatment (Vitelli, 2020). These instances may allow wandering thoughts to lead to or develop into more severe mental health issues, including anxiety, depression, substance abuse, and other psychiatric disorders. Sadly, people with pre-existing mental health disorders, vulnerable populations, and those living in coronavirus hotspots are at a higher risk of experiencing these symptoms (Sher, 2020). Many people are unable to go to their therapist's office and must communicate through a voice or video call. The inaccessibility and inconveniences can discourage patients from seeking psychiatric care at all, especially children and teens, as they may feel more vulnerable sharing their innermost thoughts and feelings via communication devices. Parents, siblings, or strangers may listen in on conversations, and that can be very unsettling for those who need mental health advice. Likewise, many patients are stuck with having to attend emergency hospital departments to receive the help they need. Although there is still access to online mental health services, there is an increased demand for this service, which makes it more difficult for people to be successful in receiving this treatment. This leaves many people who are intensely struggling with fewer resources to rely on, which, consequently, may increase the possibility of suicide (Vitelli, 2020).

Despite the damage that COVID-19 inflicts on mental health, robots may provide solutions and treatments for those who are suffering. Many robots are programmed to facilitate video calls. To explain, someone on another device can wirelessly connect to the robot and broadcast their face onto the robot's screen, which essentially provides the patient with an experience similar to traditional face-to-face contact. Throughout the last few years, many researchers developed ways to program robots so that they can act as robot companions in domestic environments. They can be programmed to act as socially assistive resources that focus on motivating, coaching, and connecting to their users (Coşar et al., 2020). Although using robots might not be ideal, it nonetheless provides some resources and relief for those who have no choice but to be in isolation during the pandemic. This may help them feel less isolated, and they can develop some sort of companionship with something in their quarantine experience.

In conclusion, robots provide countless new opportunities to help those living in isolation and long-term care homes. They assist with the enforcement of quarantine measures and provide delivery systems, opportunities to work from home, assistance to elders, and psychological support. As many of these robots are created amid the pandemic, they may not be useful to countries that are beginning to gain control of the coronavirus transmission. However, these developing robots could be practical for use in any possible future pandemics (Murphy et al., 2020). Robots have exceptional potential to evolve in ways to help people across the world. In a partnership between China's Huazhong University of Science and Technology (HUST) and the University of Toronto's Faculty of Applied Science & Engineering and the AGE-WELL Network of Centres of Excellence, scientists intend to open a research and commercialization centre specifically for robots dedicated to the care of seniors. Following the eradication of the COVID-19 pandemic, the two groups hope to expand the development worldwide and offer student and faculty research opportunities and exchanges (Do, 2020). Initiatives like these are critical for the future – we must be prepared with these robots and other resources so that we can quickly control any similar situation in the future before it gets out of hand.

Chapter 12

CONCLUSION

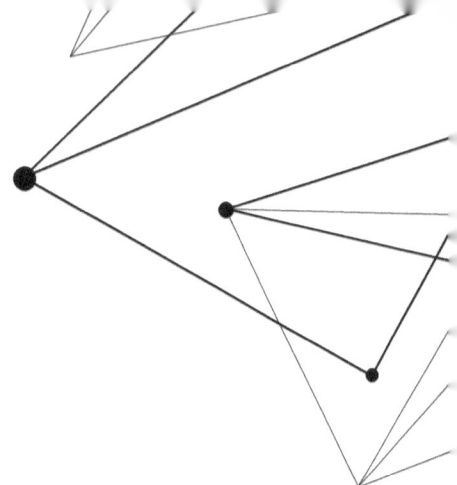

A brief glance at my calendar informed me of today's date: August 13th, 2020. Today marks the end of the fifth month since we first began quarantine; since the entire world went into lockdown. Nearly half of the year has passed, yet the months seem to have flown by in a blink of an eye. It's almost as if time itself has faltered to make way for the boundless rampage of this coronavirus storm. While my mind filtered through the series of events leading up to this point, I rolled out of bed and grabbed something to eat before finally settling down at my desk to check my emails for the day. My usual morning routine, however, has welcomed a new addition over these past five months. Instead of turning off my laptop, I proceeded to the New York Times website to catch up on the latest COVID-19 updates, closely monitoring the disease progression and reviewing federal responses. The initial shock of seeing daily increases in COVID-related deaths has long faded into a dull wariness; one that sees us mindlessly droning through our new lives for the indefinite future.

Sanitary items that were once an afterthought, such as hand sanitizers and disinfectant wipes, are now essential items in most households. Face masks also made their debut in the fashion world as the hottest new wardrobe staple. From casual strollers to high-fashion models, people in the streets adorn masks of all styles, with varying colours and designs. Whether it is a simple surgical mask, a DIY cloth mask, or a sleek polyester mask that ties together the perfect outfit, at the end of the day, we are still dreadfully aware of the real reason behind needing to wear a mask at all. Of course, social distancing is another unmistakable addition to the new norm. At every public facility, evenly spaced markers decorate the sidewalks leading up to the entrance, safely guiding visitors to social distance as they wait in line. Inside grocery stores, similar markers can be spotted near the cash registers, neatly pasted before newly mounted glass barriers that shield workers from customers. As provinces across Canada enter the final stages of reopening, these changes to our everyday lives will become more and more prevalent, before finally assembling into a picturesque demonstration of the "new normal" in a society plagued by an infectious disease.

By now, most of us have gained a clearer insight into the scope of this pandemic and the multifaceted implications it has on our society, whether it is from a medical, political, or economic standpoint. Many countries across the globe with research-intensive institutes, including Canada, joined in on the race to find a cure for COVID-19. While research has revealed similarities between the SARS-CoV-2 virus and previously widespread coronaviruses, scientists have yet to successfully develop a vaccine or treatment for the disease. Luckily, with the astounding progression of modern-day technology, we can soon accelerate the long-winded process of vaccine development. So what exactly is there to worry? Well, here is the catch: even if a COVID-19 vaccine enters the market within the next year, it would still take months, or perhaps years, before companies produce enough vaccines to distribute to the entire population and even more time before the actual distribution process is complete. Unless we want a second wave of infection to overrun the country after we just began our slow recovery to a socially and economically functioning society, we need to continue practicing the same preventative measures that have got us this far. It will likely take another few years before we can truly part with social distancing, mask-wearing, disinfecting, and sanitizing. In short, COVID-19 is not going away anytime soon, and we—as a society—need to accept that.

Our ability to regain motivation amidst prolonged adversity and to bounce back from extreme levels of acute stress perfectly unveils the beauty of human resilience. The first step to building resilience is to accept our current circumstances and identify the root causes of anxiety. For many of us, it stems from the feeling of being out of control. Starting with a global epidemic, then a political standstill, then an economic collapse; in just a few months' time, we were thrown into a whirlpool of crises without any idea on how to get out. This is when the fear strikes, knowing that we do not have the power to change anything or to escape. Only by acknowledging our misfortunes and accepting this new way of life would we be able to settle on the reality of the present moment and regain our footing. However, adversity will not dissipate with mere acknowledgement. Once we have solidified the first step, we must set out the next step to shift the balance.

Now that some resemblance of normalcy has been re-established, it is time for us to reevaluate our options, understand what we can and cannot control, and take action to control all that we can. As residents of the 21st century, we are fortunate enough to grow alongside the exponential advancements in science and technology, with first-hand access to all the newest gadgets and discoveries. Robots and artificial intelligence are two of the most avant-garde inventions to have emerged from the information age and are already prevalent in everyday life. Over the past several decades, the field of robotics has extended a helping hand to drastically different domains of mankind, completing tasks as simple as vacuuming kitchen floors to fulfilling outer space missions and collecting geographic samples from Mars. In the context of COVID-19, robotics holds the most potential for enhancing the performance of healthcare sectors, providing healthcare professionals with a tool to control clinical environments amid the rush of this pandemic-induced chaos. The key to regaining control, therefore, lies in our ability to maximize the potential of these technological devices.

One of the few things which we can control when it comes to COVID-19 is containing the spread of disease by reducing transmission among ourselves. Although authorities already implemented social distancing and warned against person-to-person contact, the effectiveness of these policies is compromised by the lack of compliance from the general public. Hospitals and other clinical environments are areas with the highest risks of exposure to the virus, which is where the role of robots emerges. To begin, robot-assisted surgery (RAS) opens the door for many new opportunities and advancements in the medical field. It is a practical alternative for medical professionals to utilize as it limits human exposure, reduces one's hospital stay, increases hospital turnover, and protects people from virus transmission. Robots also provide methods of disinfection in hospitals that increase the efficiency of cleaning when compared to human cleaners. Likewise, they provide significant relief for the current burden on healthcare workers. In hospitals overloaded with coronavirus patients, robots can be leveraged to screen visitors at entrances, monitor vital signs, deliver food and medicine, and possibly provide mental health support for patients. AI-based robots also create an opportunity for the incorporation of telemedicine for doctors to provide quality medical care without putting anyone at risk of infection. The use of robots in the hospital offers significant advantages that optimize a hospital's efficiency and effectiveness during troubling times.

Knowing that COVID-19 will not subside anytime soon, we must use as many resources as possible to develop a multitude of prevention and screening methods—another component of disease management that we can control. Artificial intelligence is an excellent resource for such applications as it demonstrates the potential to reduce the amount of time to receive testing results by creating infrared scanning systems, efficient contact-tracing, and mass screening. Local facilities may also utilize AI to scan public areas to monitor the maintenance of physical distancing practices and mask-wearing. If select individuals fail to meet these requirements, the AI systems may then set off an alarm to notify authorities or business-owners to intervene. Companies are also creating similar AI-based databases and prevention methods to keep the public informed. Through this, we can see how artificial intelligence is a vital resource in minimizing risks and viral transmission, altogether creating a safer environment for us to live in.

That said, not only do robots play a crucial role in hospital settings, but they also provide many alternatives during quarantine and in nursing homes. They can deliver essential items and medications to people in quarantine, provide options to work remotely, and ease the burden on people living in long-term care homes. Robots provide remote healthcare to those living in quarantine as it reduces the number of people that need to leave their homes to meet with a healthcare professional. Many workers also use robots and technology at home during the pandemic, which minimizes their time spent outside and further reduces the risk of transmission. Throughout the world, there have been many coronavirus outbreaks in long-term care homes, making it all the more important for us to pay extra attention to our elders. With a shortage of nursing personnel in light of current circumstances, robots can fill those gaps by providing additional support. They offer physical and psychological assistance and even help elders to keep in contact with their families. Altogether, robots provide manifold opportunities and resources to those living in isolation and long-term care homes, allowing individuals to preserve their sense of self during these times of uncertainty.

Through the previous examples, we can see how robots might contribute to the conversion of doubt to belief in the face of adversity. This marks a decisive shift in balance, which leads us to the final step of building resilience: perseverance. Aside from COVID-19, another well-known pandemic of modern history was the 1918 Spanish Flu. Much like the coronavirus, the Spanish Flu took the world by storm and soon became the most challenging medical mystery at the time to doctors and scientists. Authorities enforced lockdown procedures in urban areas, hospitals overflooded with infected patients and a shortage of personnel, and people followed instructions to remain indoors and avoid contact. Even with maximal efforts, the Spanish Flu still left an indelible mark on society, wiping out families and eradicating economies. In its wake, those that were infected either succumbed to illness or developed immunity; two extremes with no in-between. To make matters worse, recovery from the Spanish Flu was shortly succeeded by the eruption of World War I, yet despite the deadliest of circumstances, the world successfully emerged from its ashes in a resilient showcase of societal rebirth. This is what it means to persevere. If we have done it once in the past, with science and technology that are hardly comparable to what exists today, then we have the potential to persevere just as well through the pandemic of today.

So what now? What might the future hold for us? While we certainly do not have complete control over what life throws at us, we can do our best to prepare for similar scenarios. Scientific manifestations such as AI will continue to progress and become more intelligent, and thus can be applied to a wider variety of contexts to suit our needs; particularly in the case of another pandemic. From an epidemiological perspective, they could analyze data sets at blazing speeds to predict the geographical spread of disease. More AI-based robots in everyday situations such as healthcare and retail would indicate fewer chances of dispersing pathogens through biological mechanisms such as coughing and sneezing. Of course, this would mean that the robots would have to replace the humans who actively distribute these viral particles. But, as with most things in life, there is always a price to pay, whether it be directly or indirectly. The only question that remains is whether this is a price we are willing to pay, and if not, how can we still reap the benefits of robotics while mitigating its risks?

References

1. Cucinotta D, Vanelli M. WHO Declares COVID-19 a Pandemic. Acta Biomed [Internet]. 2020 [cited 2020 July 16];91(1):157-160. Available from: http://10.23750/abm.v91i1.9397

2. Madhav N, Oppenheim B, Gallivan M, Mulembakani P, Rubin E, Wolfe N. Pandemics: Risks, Impacts, and Mitigation. In: Jamison DT, Gelband H, Horton S, Prabhat J, Ramanan L, Mock CN, Nugent R, editors. Disease Control Priorities: Improving Health and Reducing Poverty. 3rd edition [Internet]. Washington (DC): The International Bank for Reconstruction and Development / The World Bank; 2017 [cited 2020 July 23]. Chapter 17. Available from: https://www.ncbi.nlm.nih.gov/books/NBK525302/

3. Wu Y-C, Chen C-S, Chan Y-J. The outbreak of COVID-19: An overview. Journal of the Chinese Medical Association [Internet]. 2020 [cited 2020 July 23];83(3):217–20. Available from: https://dx.doi.org/10.1097%2FJCMA.0000000000000270

4. World Health Organization: WHO [Internet]. Switzerland: WHO; 2020. Coronavirus disease (COVID-19); 2020 July 13 [cited 2020 July 13]. Available from: https://www.who.int/emergencies/diseases/novel-coronavirus-2019

5. COVID-19 to slash global economic output by $8.5 trillion over next two years [Internet]. United States: United Nations; 2020 [cited 2020 July 17]. Available from: https://www.un.org/development/desa/en/news/policy/wesp-mid-2020-report.html

6. Peterson S, Mesley W. COVID-19 blame game intensifies with online ad campaigns paid for by China and the U.S [Internet]. Canada: CBC; 2020 [cited 2020 July 23]. Available from: https://www.cbc.ca/news/world/coronavirus-covid-china-united-states-political-campaign-1.5555245

7. Canada.ca [Internet]. Canada: Government of Canada; 2018. Canadarm - Canadian Space Agency; 21 June 2018 [cited 2020 July 24]. Available from: https://www.asc-csa.gc.ca/eng/canadarm/default.asp

8. Liu, X, Faes L, Kale AU, Wagner SK, Fu DJ, Bruynseels A, et al. A comparison of deep learning performance against health-care professionals in detecting diseases from medical imaging: a systematic review. Lancet Digital Health [Internet]. 2019 [cited 2020 July 23];1(6):e271-e297. Available from: https://doi.org/10.1016/ S2589-7500(19)30123-2

9. Koonin, EV. The wonder world of microbial viruses. Expert Rev Anti Infect Ther [Internet]. 2010 [cited 2020 July 23];8(10):1097-1099. Available from: https://dx.doi.org/10.1586%2Feri.10.96

10. 10. Wu Y-C, Chen C-S, Chan Y-J. The outbreak of COVID-19. Journal of the Chinese Medical Association. 2020;83(3):217–20.

11. Blewett T. Tracking the coronavirus, from Wuhan, China to Canada's capital: A COVID-19 timeline [Internet]. Ottawa Citizen. Ottawa Citizen; 2020 [cited 2020Jul18]. Available from: https://ottawacitizen.com/news/local-news/tracking-the-coronavirus-from-wuhan-china-to-canadas-capital-a-covid-19-timeline

12. Ciotti M, Angeletti S, Minieri M, Giovannetti M, Benvenuto D, Pascarella S, et al. COVID-19 Outbreak: An Overview. Chemotherapy. 2019;64(5-6):215–23.

13. Wilder-Smith A, Chiew CJ, Lee VJ. Can we contain the COVID-19 outbreak with the same measures as for SARS? The Lancet Infectious Diseases. 2020;20(5).

14. Ye Z-W, Yuan S, Yuen K-S, Fung S-Y, Chan C-P, Jin D-Y. Zoonotic origins of human coronaviruses. International Journal of Biological Sciences. 2020;16(10):1686–97.

15. Chan JF-W, Kok K-H, Zhu Z, Chu H, To KK-W, Yuan S, et al. Genomic characterization of the 2019 novel human-pathogenic coronavirus isolated from a patient with atypical pneumonia after visiting Wuhan. Emerging Microbes & Infections. 2020;9(1):221–36.

16. Q&A on coronaviruses (COVID-19) [Internet]. World Health Organization. World Health Organization; 2020 [cited 2020Jul18]. Available from: https://www.who.int/emergencies/diseases/novel-coronavirus-2019/question-and-answers-hub/q-a-detail/q-a-coronaviruses

17. Petersen E, Koopmans M, Go U, Hamer DH, Petrosillo N, Castelli F, et al. Comparing SARS-CoV-2 with SARS-CoV and influenza pandemics. The Lancet Infectious Diseases. 2020;

18. 18. Canada PHAof. Government of Canada [Internet]. Canada.ca. 2020 [cited 2020Jul20]. Available from: https://www.canada.ca/en/public-health/services/publications/diseases-conditions/social-distancing.html

19. Lyu W, Wehby GL. Community Use Of Face Masks And COVID-19: Evidence From A Natural Experiment Of State Mandates In The US. Health Affairs. 2020;

20. Chung E. Mandatory mask laws are spreading in Canada | CBC News [Internet]. CBCnews. CBC/Radio Canada; 2020 [cited 2020Aug11]. Available from: https://www.cbc.ca/news/health/mandatory-masks-1.5615728

21. "Solidarity" clinical trial for COVID-19 treatments [Internet]. World Health Organization. World Health Organization; 2020 [cited 2020Jul20]. Available from: https://www.who.int/emergencies/diseases/novel-coronavirus-2019/global-research-on-novel-coronavirus-2019-ncov/solidarity-clinical-trial-for-covid-19-treatments

22. Testing for COVID-19 [Internet]. Centers for Disease Control and Prevention. Centers for Disease Control and Prevention; 2020 [cited 2020Jul20]. Available from: https://www.cdc.gov/coronavirus/2019-ncov/symptoms-testing/testing.html

23. Cascella M, Rajnik M, Cuomo A, Dulebohn SC, Di Napoli R. Features, Evaluation and Treatment Coronavirus (COVID-19). StatPearls [Internet]. Treasure Island (FL): StatPearls Publishing; 2020. [cited 2020 July 16]. Available from: https://www.ncbi.nlm.nih.gov/books/NBK554776/

24. Canada.ca [Internet]. Canada: Government of Canada; 2020. Coronavirus disease (COVID-19): Outbreak update; 2020 July 13 [cited 2020 July 13]. Available from: https://www.canada.ca/en/public-health/services/diseases/2019-novel-coronavirus-infection.html

25. World Health Organization: WHO [Internet]. Switzerland: WHO; 2020. Coronavirus disease (COVID-19); 2020 July 13 [cited 2020 July 13]. Available from: https://www.who.int/emergencies/diseases/novel-coronavirus-2019

26. Cucinotta D, Vanelli M. WHO Declares COVID-19 a Pandemic. Acta Biomed [Internet]. 2020 [cited 2020 July 16];91(1):157-160. Available from: http://10.23750/abm.v91i1.9397

27. Kenyon C. Flattening-the-curve associated with reduced COVID-19 case fatality rates- an ecological analysis of 65 countries. J Infect [Internet]. 2020 [cited 2020 July 16];81(1):e98-e99. Available from: https://dx.doi.org/10.1016%2Fj.jinf.2020.04.007

28. Cavallo J, Donoho D, Forman H. Hospital Capacity and Operations in the Coronavirus Disease 2019 (COVID-19) Pandemic—Planning for the Nth Patient. JAMA Health Forum [Internet]. 2020 [cited 2020 July 17];1(3):e200345. Available from: https://jamanetwork.com/channels/health-forum/fullarticle/2763353

29. Armocida B, Formenti B, Ussai S, Palestra F, Missoni E. The Italian health system and the COVID-19 challenge. Lancet Public Health [Internet]. 2020 [cited 2020 July 16];5(5):e253. Available from: https://dx.doi.org/10.1016%2FS2468-2667(20)30074-8

30. Ho S. Tweets comparing U.S. and Canadian health care during pandemic strike nerve [Internet]. Canada: Bell Media; 2020 [cited 2020 July 16]. Available from: https://www.ctvnews.ca/health/coronavirus/tweets-comparing-u-s-and-canadian-health-care-during-pandemic-strike-nerve-1.5001456

31. Holroyd-Leduc J, Laupacis A. Continuing care and COVID-19: a Canadian tragedy that must not be allowed to happen again. CMAJ [Internet]. 2020 [cited 2020 July 17];192(23):E632-E633. Available from: https://doi.org/10.1503/cmaj.201017

32. Imbert F, Stevens P, Fitzgerald M. Stock market live Tuesday: Dow drops 410 points, down 23% in 2020, Worst first quarter ever [Internet]. United States: CNBC; 2020 [cited 2020 July 16]. Available from: https://www.cnbc.com/2020/03/31/stock-market-today-live.html

33. COVID-19 to slash global economic output by $8.5 trillion over next two years [Internet]. United

States: United Nations; 2020 [cited 2020 July 17]. Available from: https://www.un.org/development/desa/en/news/policy/wesp-mid-2020-report.html

34. Canada lost a record one million jobs in March [Internet]. United Kingdom: BBC News; 2020 [cited 2020 July 16]. Available from: https://www.bbc.com/news/world-us-canada-52232674

35. Patel J, Nielsen F, Badiani A, Assi S, Unadkat V, Patel B, Ravindrane R, Wardle H. Poverty, inequality and COVID-19: the forgotten vulnerable. Public Health [Internet]. 2020 [cited 2020 July 17];183:110-111. Available from: https://dx.doi.org/10.1016%2Fj.puhe.2020.05.006

36. CTV News. COVID-19 survivor receives US$1.1M hospital bill [Internet]. Canada: Bell Media; 2020 [cited 20 July 17]. Available from: https://www.ctvnews.ca/health/coronavirus/covid-19-survivor-receives-us-1-1m-hospital-bill-1.4983547

37. COVID-19 in Racial and Ethnic Minority Groups [Internet]. United States: Centers for Disease Control and Prevention; 2020 [cited 2020 July 17]. Available from: https://www.cdc.gov/coronavirus/2019-ncov/need-extra-precautions/racial-ethnic-minorities.html

38. Judd A. Anti-Asian hate crimes: 29 cases in Vancouver so far this year, compared to 4 last year [Internet]. Canada: Corus Entertainment; 2020 [cited 2020 July 17]. Available from: https://globalnews.ca/news/6974373/vancouver-hate-crimes-coronavirus/

39. Chen H, Trinh J, Yang G. Anti-Asian sentiment in the United States – COVID-19 and history. Am J Surg [Internet]. 2020 [cited 2020 July 17]. Available from: https://dx.doi.org/10.1016%2Fj.amjsurg.2020.05.020

40. Chen H, Trinh J, Yang G. Anti-Asian sentiment in the United States – COVID-19 and history. Am J Surg [Internet]. 2020 [cited 2020 July 17]. Available from: https://dx.doi.org/10.1016%2Fj.amjsurg.2020.05.020

41. Lee B. Trump Once Again Calls Covid-19 Coronavirus The 'Kung Flu' [Internet]. United States: Forbes Media; 2020 [cited 2020 July 17]. Available from: https://www.forbes.com/sites/brucelee/2020/06/24/trump-once-again-calls-covid-19-coronavirus-the-kung-flu/#6fee2b791f59

42. Reger M, Stanley I, Joiner T. Suicide Mortality and Coronavirus Disease 2019—A Perfect Storm?. JAMA Psychiatry [Internet]. 2020 [cited 2020 July 17]. Available from: http://jamanetwork.com/article.aspx?doi=10.1001/jamapsychiatry.2020.1060

43. Yang G-Z, Nelson BJ, Murphy RR, Choset H, Christensen H, Collins SH, et al. Combating COVID-19—The role of robotics in managing public health and infectious diseases. Science Robotics. 2020;5(40).

44. Cousins B. Lack of resources led to limited COVID-19 testing, but new options are on the way [Internet]. Coronavirus. CTV News; 2020 [cited 2020Aug11]. Available from: https://www.ctvnews.ca/health/coronavirus/lack-of-resources-led-to-limited-covid-19-testing-but-new-options-are-on-the-

way-1.4891161

45. Imran A, Posokhova I, Qureshi HN, Masood U, Riaz S, Ali K, et al. AI4COVID-19: AI enabled preliminary diagnosis for COVID-19 from cough samples via an app. Informatics in Medicine Unlocked. 2020Jun26;:100378.

46. Lee OE, Davis B. Adapting 'Sunshine,' A Socially Assistive Chat Robot for Older Adults with Cognitive Impairment: A Pilot Study. Journal of Gerontological Social Work. 2020;:1–3.

47. Arthur C, Ruan S. In China, robot delivery vehicles deployed to help with COVID-19 emergency [Internet]. UNIDO. United Nations Industrial Development Organization; 2020 [cited 2020Aug11]. Available from: https://www.unido.org/stories/china-robot-delivery-vehicles-deployed-help-covid-19-emergency

48. Strickland A. How robots could help us combat pandemics in the future [Internet]. CNN. Cable News Network; 2020 [cited 2020Aug11]. Available from: https://www.cnn.com/2020/03/25/world/robots-pandemics-coronavirus-scn/index.html

49. Jain N. How Trust In Robots Can Help Us Fight The Next Pandemic [Internet]. Forbes. Forbes Magazine; 2020 [cited 2020Aug11]. Available from: https://www.forbes.com/sites/neerajain/2020/05/20/how-trust-in-robots-can-help-us-fight-the-next-pandemic/

50. Statistics Canada May 8, 2020. Available from: https://www150.statcan.gc.ca/n1/daily-quotidien/200508/dq200508a-eng.htm

51. U.S. BUREAU OF LABOR STATISTICS May 8, 2020. Available from: https://www.bls.gov/news.release/archives/empsit_05082020.htm

52. Frey CB, Osborne MA. The future of employment: How susceptible are jobs to computerisation? Technological Forecasting and Social Change. 2017Jan;114:254–80.

www.ingramcontent.com/pod-product-compliance
Lightning Source LLC
Chambersburg PA
CBHW022109160426
43198CB00008B/413